Arduino Oscilloscope Projects

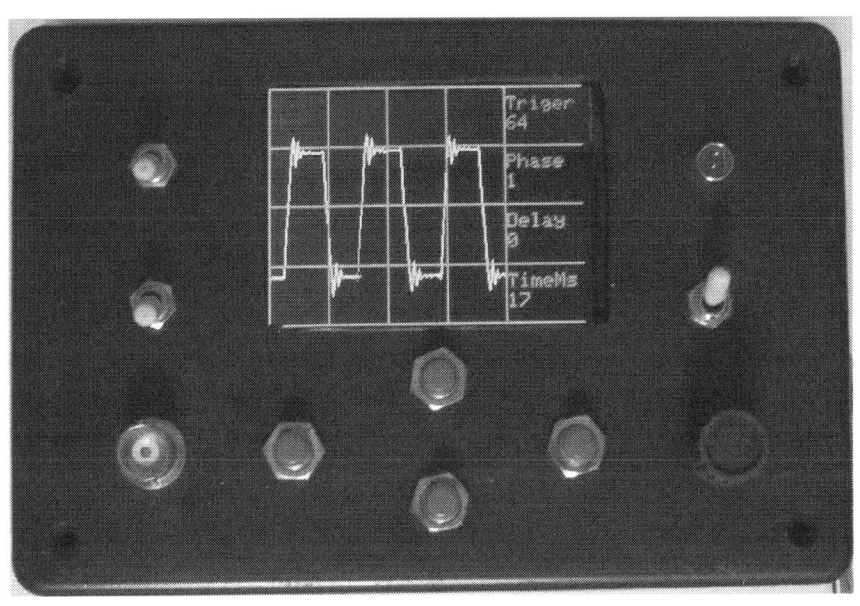

Robert J Davis II

Arduino Oscilloscope Projects

Copyright 2015 by Robert J Davis II

All rights reserved

In essence this is a continuation of my previous book "Arduino LCD Projects" where I introduced many LCD screens and the CA3306 fast analog to digital converter to make a fast 5 MSPS oscilloscope. A magazine article about the Arduino and CA3306 based oscilloscope created so much interest that I decided to write a book showing some of the most popular analog to digital converters and some of the most popular LCD screen's together in Oscilloscope, and logic analyzer, types of applications. Some chapters of this book repeat some of the designs that were found in the "Arduino LCD Projects" book.

Years ago, as a teenager, I tried many times to build an oscilloscope. However I never had a working oscilloscope until I built one as part of my CIE training. Then in my book "Digital and Computer Projects" I introduced the concept of a parallel port PC based oscilloscope. I even designed a circuit board for it. However the parallel port is now history and USB is where things are at. Along came the Arduino and it is the basis for a simple USB oscilloscope.

There are many fast analog to digital converters that I want to cover in this book. There are some fast analog to digital converters that can operate at speeds of 50 MSPS (Million Samples Per Second) or even faster. Since the Arduino Uno is not nearly that fast, those analog to digital converters must be used with a FIFO (Fist In First Out) high speed memory buffer,

The use of fast analog to digital converters with an LCD screen to make an oscilloscope is an area that raises a lot of design questions. I hope to answer many of your questions in this book.

Once again the safe construction and operation of these devices is solely up to the reader. The reader must take all safety precautions and accept responsibility for the safe operation of these devices. There are no guarantees implied with the circuit designs and software programs that are found within this book.

The most important thing is to have fun! Try out some circuits and see what you like. Make your own hardware and software improvements. I am sure you can come up with better designs!

Table of Contents.

1. Overview of Arduino Uno Oscilloscopes.................. 4
    Internal analog to digital converter
    External analog to digital converter
    FIFO and External analog to digital converter

2. Analog Input Section Design................................ 8
    TL082 or LF353
    AD744 and NE5532

3. Fast Analog to Digital Converters.......................... 16
    CA3306
    CA3318
    AD775
    AD8703
    TLC5510
    Synchronizing the Clocks

4. Monochrome LCD - QC12864B............................. 25
    Internal Analog to Digital Converter
    Six Channel Logic Analyzer
    External Analog to Digital Converter

5. Serial Color LCD - 1.8TFT SPI 44
    Internal Analog to Digital Converter
    External Analog to Digital Converter

6. Parallel Color LCD - TFT240_262K........................ 57
    Internal Analog to Digital Converter
    Six Channel Logic Analyzer
    External Analog to Digital Converter

7. Serial Color LCD - 2.2 to 2.8 SPI........................... 75
    Internal Analog to Digital Converter
    External Analog to Digital Converter

8. Putting the Oscilloscope into a Case 89

Bibliography.. 93

Chapter 1

Overview of Arduino Oscilloscopes

There are basically three types of Arduino Uno based oscilloscopes. First you can use the built in analog to digital converter but modify its timing to make it much faster. Secondly you can add a fast external analog to digital converter with either 6 or 8 bits of accuracy depending on if you use port C or port D. The third type uses a FIFO (First in First Out) buffer to sample data at speeds faster than the Arduino can operate at.

This is the block diagram of the first type of oscilloscope. For this type of oscilloscope you typically apply the buffered signal to A0. With software modifications to the clock timing setup of the built in analog to digital converter you can achieve 100,000 samples per second or .1 MSPS.

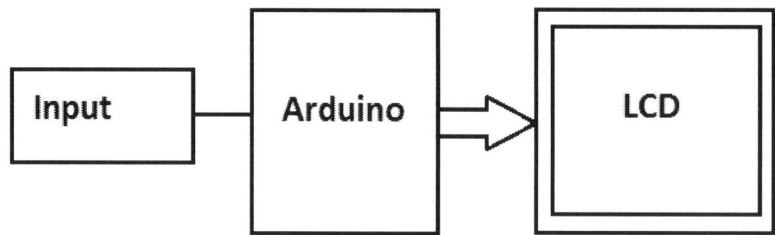

Here is the code to change the clock settings in order to speed up the Arduino's internal analog to digital converter by a factor of almost 16. The change will reduce the accuracy of the converter. This code will clear the most significant bit of the three timing control bits. To do that, add this code right after the "void loop() {".

```
// Clear bit 2 of ADC pre-scalier from 125KHz to 2 MHz
   ADCSRA &= ~(1 << ADPS2);
```

For the next type of oscilloscope we add an analog to digital converter and use the parallel input command for speeds up to 5 million samples per second. The PINC (Parallel Input Port C) command uses A0 to A5. The PIND (Parallel Input Port D) command uses D0 to D7.

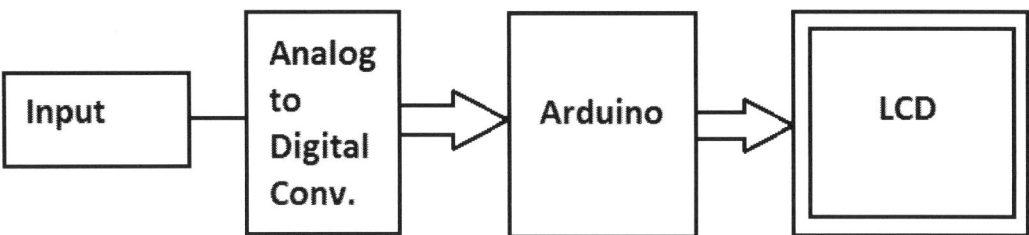

There are two variations to the external analog to digital converter setup. The first uses A0 to A5 and is hence it is limited to only six bits. This is an ideal setup for the CA3306 as it is a six bit converter. You can use an eight bit converter but only connect the top six outputs. That is to say D7 becomes D5, D6 becomes D4, etc. To rapidly read the contents of the six analog inputs the "PINC" command is used.

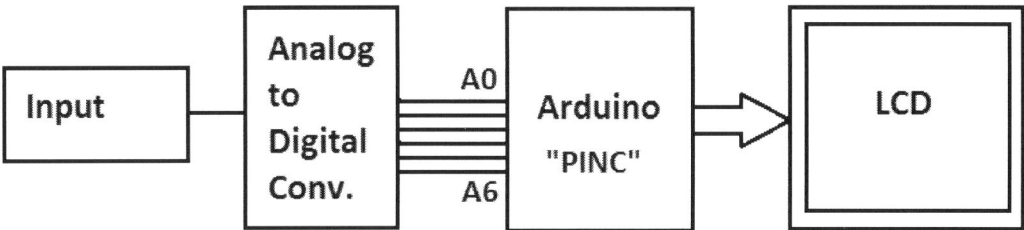

The second variation of the external analog to digital converter setup uses the D0 to D7 pins. They are read rapidly by using the "PIND" command. However, D0 and D1 are also used for the USB cable to talk to the Arduino. They either have to be disconnected to update the software in the Arduino, or somehow disabled.

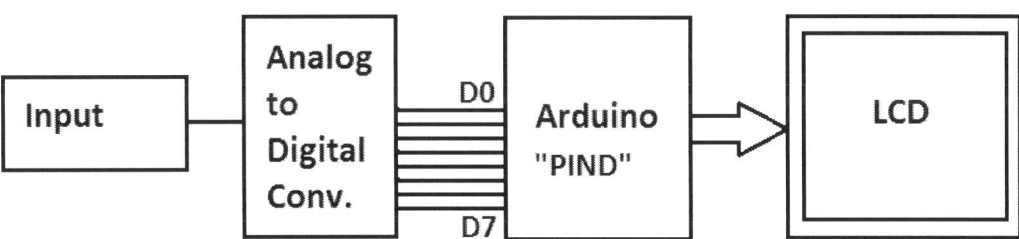

For the third type of oscilloscope they add a memory buffer that is called a "FIFO". That stands for "First In First Out". FIFO's can operate at speeds up to 100MSPS or more. This setup will also require a clock divider and select circuit to selectively slow things down or else it will always sample at the maximum speed of 20 to 100 MSPS. To be honest a FIFO setup might be a little too complicated to build on a breadboard setup. I have built it using wire wrap and a printed circuit board. A FIFO setup was included in my older book "Digital and Computer Projects"

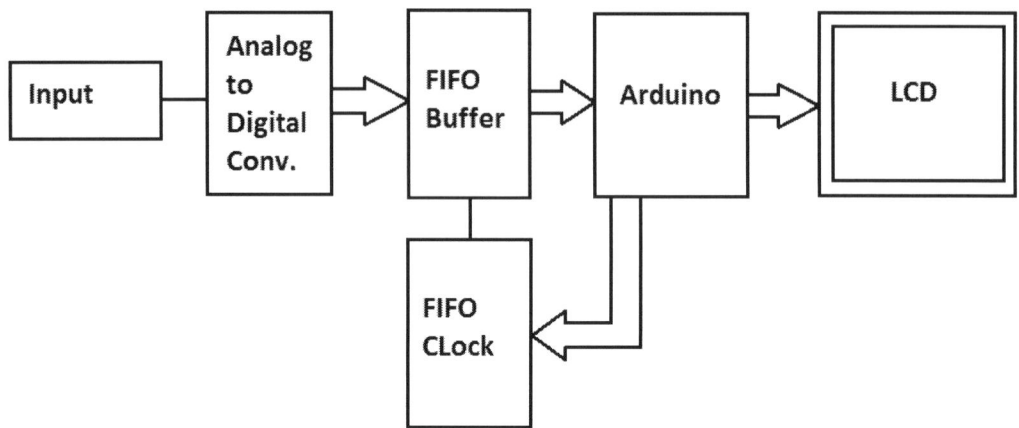

Here is a schematic of a typical FIFO setup, in a six bit arrangement.

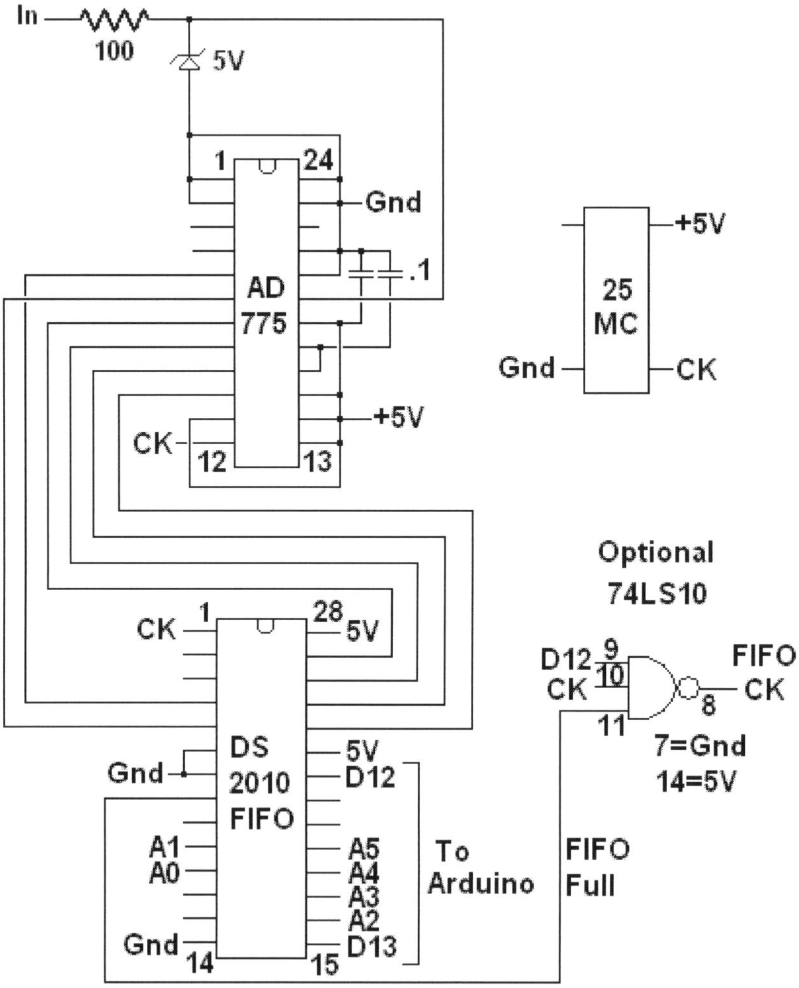

This book will concentrate primarily on the first two oscilloscope designs, as they are much easier ones to make. The designs will be using the internal analog to digital converter and using an external six or eight bit analog to digital converter. Each of the "blocks" in the first four schematic diagrams above will be covered in detail in the following chapters of this book. Those individual parts or blocks are all somewhat interchangeable.

Chapter 2

Analog Input Section Design

The input section serves three purposes. First an analog input section provides high impedance to the device under test. This prevents loading on the device under test. Secondly it provides attenuation of the input. Usually it is a 1/10 attenuation to allow inputs of up to 10 times normal to be displayed. With a 10X probe over 100 volt signals can be viewed. The analog section also provides "bias". Most analog to digital converters can only take a positive input, so the analog signal needs to be biased so that the signal falls into the middle of the converters input range.

Here is a very simple input setup schematic. It only provides bias.

```
              100              To Analog
In ──||──/\/\/\──┬──────────── to Digital
    .47 uF       │      ▽ 5V   Converter
              100K ⌇         
                   │         
                   │         
             5V──/\/\──┴──Gnd
                  10K
                Position
```

A better analog input section can provide input attenuation of 1/10 for a 10 volt signal a high impedance buffer and a bias with just one IC. You can use a FET input dual Op amplifier like either a LF353 or a TL082. The drawback of this improved input section is that you will need a positive and negative 9 to 12 volt power source. Two 9 volt batteries will work.

You could also use a 9 volt AC adapter and a 555 power inverter to provide the negative voltage. This improved input circuit will also present a smaller load, of one million ohms, to the circuit that is under test. It will also provide some selection of the input attenuation as well as the gain. On top of that it has diode over voltage protection, for the input and for the analog to digital converter.

Here is the schematic diagram of the improved analog input section.

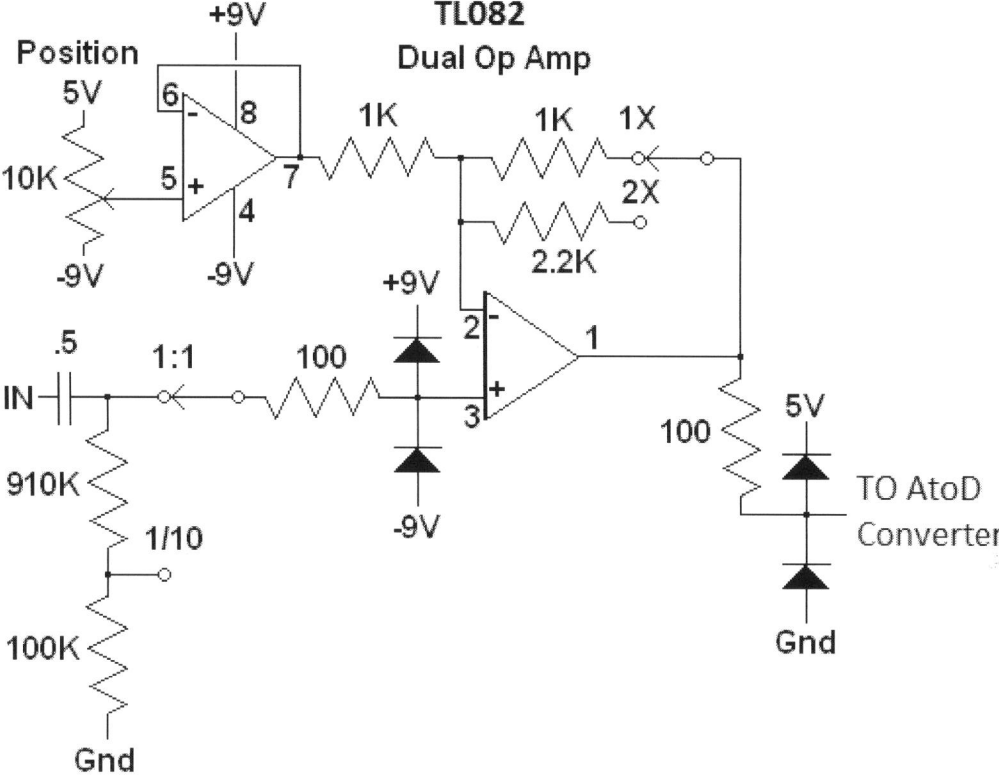

Next we will improve on the analog input section that we just saw. In the previous design when you changed the gain of the OP amplifier the bias changed requiring it to be re-adjusted. It is better to not have an adjustable gain to avoid this problem. The "gain" can also be adjusted in the software. This change will even make the analog input section simpler in design. You can use a TL082, a LF353 or many of the other FET input dual OP amplifiers.

The "bias" control provides a bias to the analog to digital converter. You will need a bias of about 1.3 volts for 2.6 volts as the maximum input. The 1.3 volts then becomes the "zero" position in the middle of the LCD screen. This bias is the "0" signal as the analog to digital converter uses 2.6 volts as its maximum input voltage and does not allow any negative voltage inputs. The bias voltage will be higher, about 2.5 volts, with the CA3306 and CA3318 as they have a higher reference voltage.

Up next is the improved schematic.

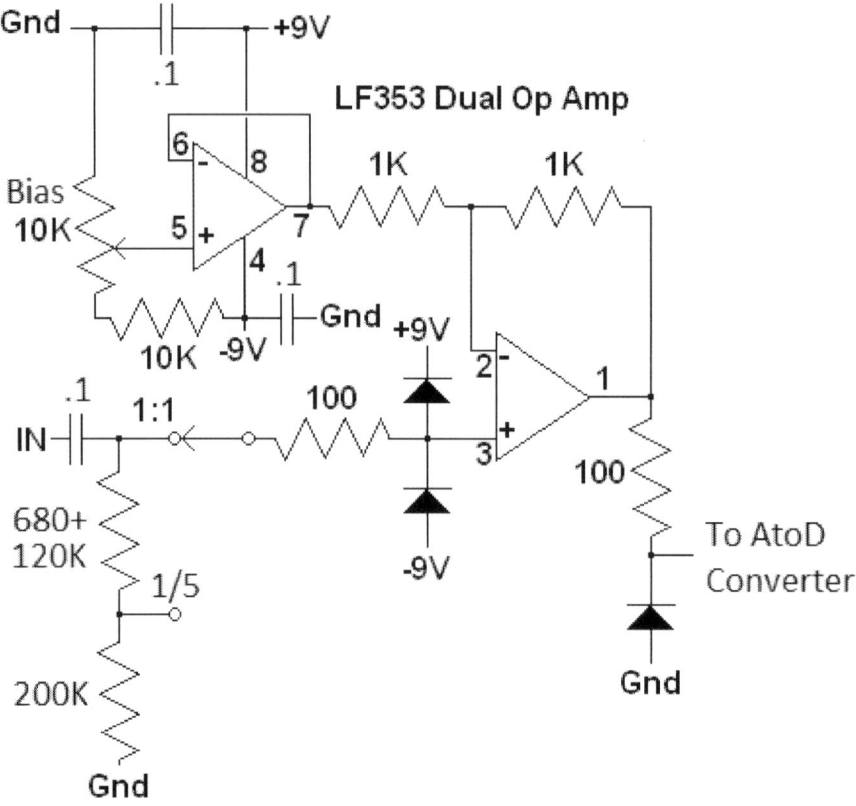

Up next is a picture of what the analog section looks like when it is wired up on a breadboard. Do not forget to use the .1 uF filter capacitors that are connected to pins 4 and 8 of the op amp to ground. These capacitors will make a huge difference in the amount of noise.

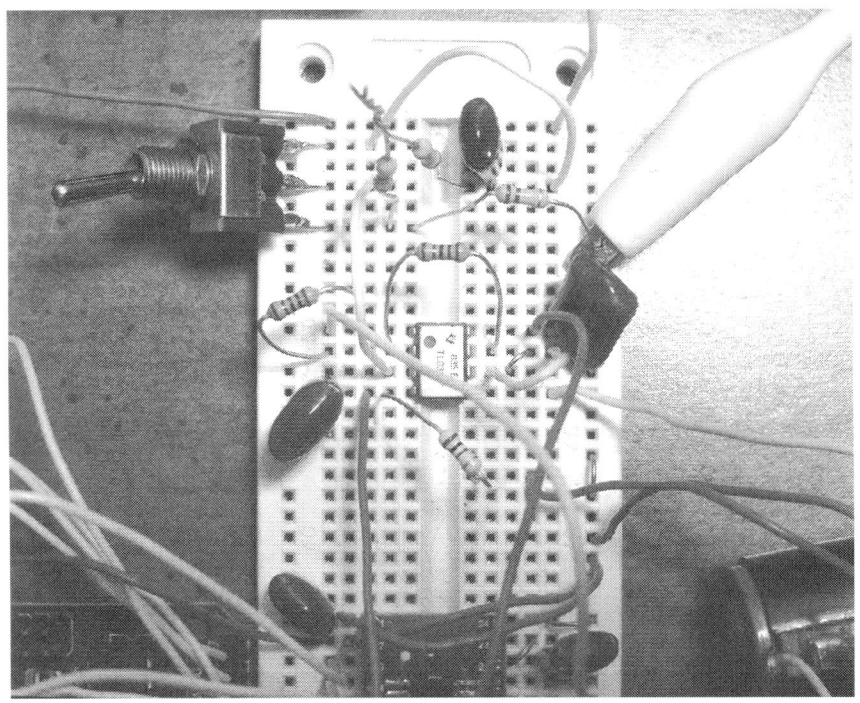

The power supply for the OP amp needs to provide two voltages. The +9 volts comes from an AC adapter that is powering the Arduino Uno. It can be tapped into via the Arduino "Vin" pin. For a -9 volt source you can either use a 9V battery or even better use a 555 oscillator with two diodes to form a simple power inverter. Here is a schematic for a 555 inverter. The diodes can be a 1N4001 or similar. Under load the output is only about -6 volts but that works fine.

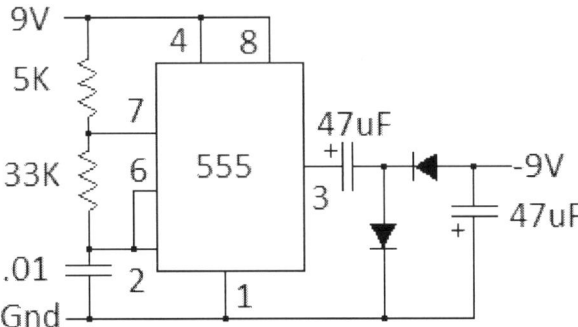

For an even sharper square wave the TL082 OP amp can be replaced with an AD744. It will require some rewiring to make it work. Up next there are two pictures that compare the two amplifiers processing a square wave at 65 KHz. The TL082 is shown in the first picture and the AD744 is in the second picture. Notice how much sharper and squarer the corners of the square wave are in the second picture.

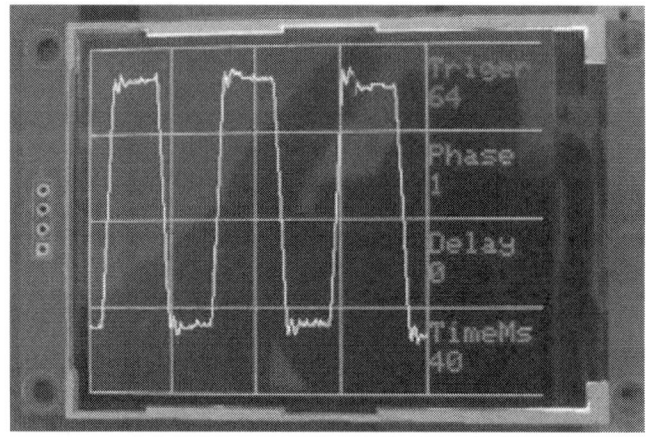

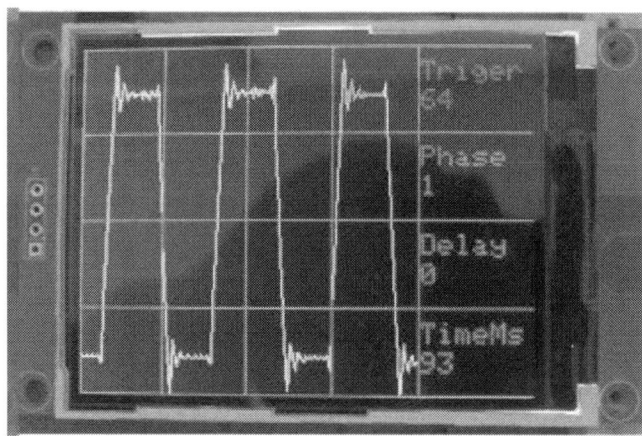

Although it can operate on less, the AD744 works best with positive and negative 9 to 12 volts. You can power the Arduino from a 9 to 12 volt AC adapter. Then use the "Vin" to power the AD744 as well as the 555 inverter to get the negative power source.

One problem with the AD744 is that it will not allow me to add the bias voltage the same way that the TL082 does. Instead I had to AC couple it and use a variable resistor to add bias directly to the input of the analog to digital converter.

Up next is the AD744 input section schematic diagram.

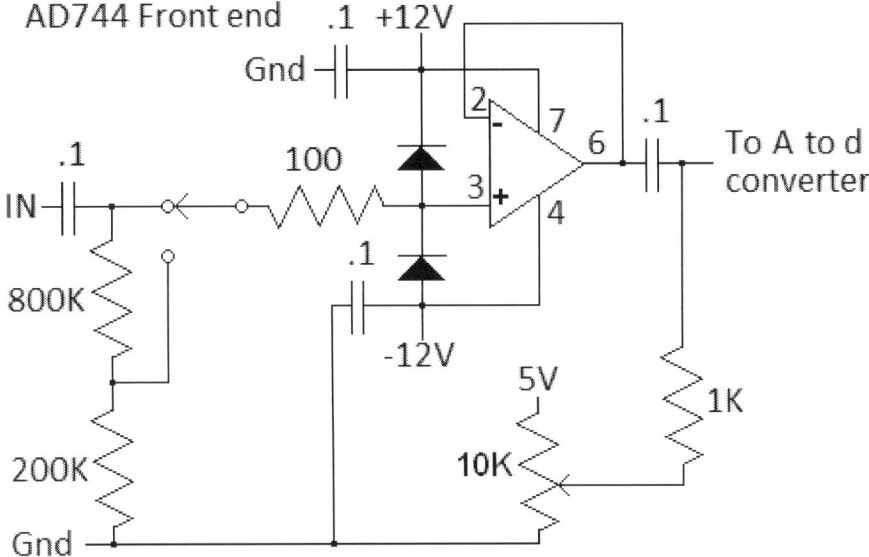

Do not forget the .1 uF capacitors from plus and minus 12 volts to ground. Also a 5pF capacitor needs to go from pin 5 to pin 8 of the AD744 or you will see a lot of noise! The protection diodes should be 1N914's. I made the 800K resistor out of a 470K and a 330K resistor in series. Some oscilloscope input designs use a 900K and 100K resistor combination to give a 1X and .1X attenuation selection.

There could be a 1-10 pF trimmer capacitor across the 800 K resistor. Also there could be a 10-100 pF trimmer capacitor across the 200K resistor. These trimmer capacitors might be needed to square up the edges of a square wave.

To better couple the AD744 to the analog to digital converter you need to add the TL082 circuit back in. But there is a better way to do that. Instead of the TL082 use a NE5532 or an AD827. Instead of topping out a 4MHz the NE5532 tops out at 10MHz and the AD827 tops out at 50MHz! They also have a much lower noise figure. So, up next is a schematic of the two circuits put together into one schematic.

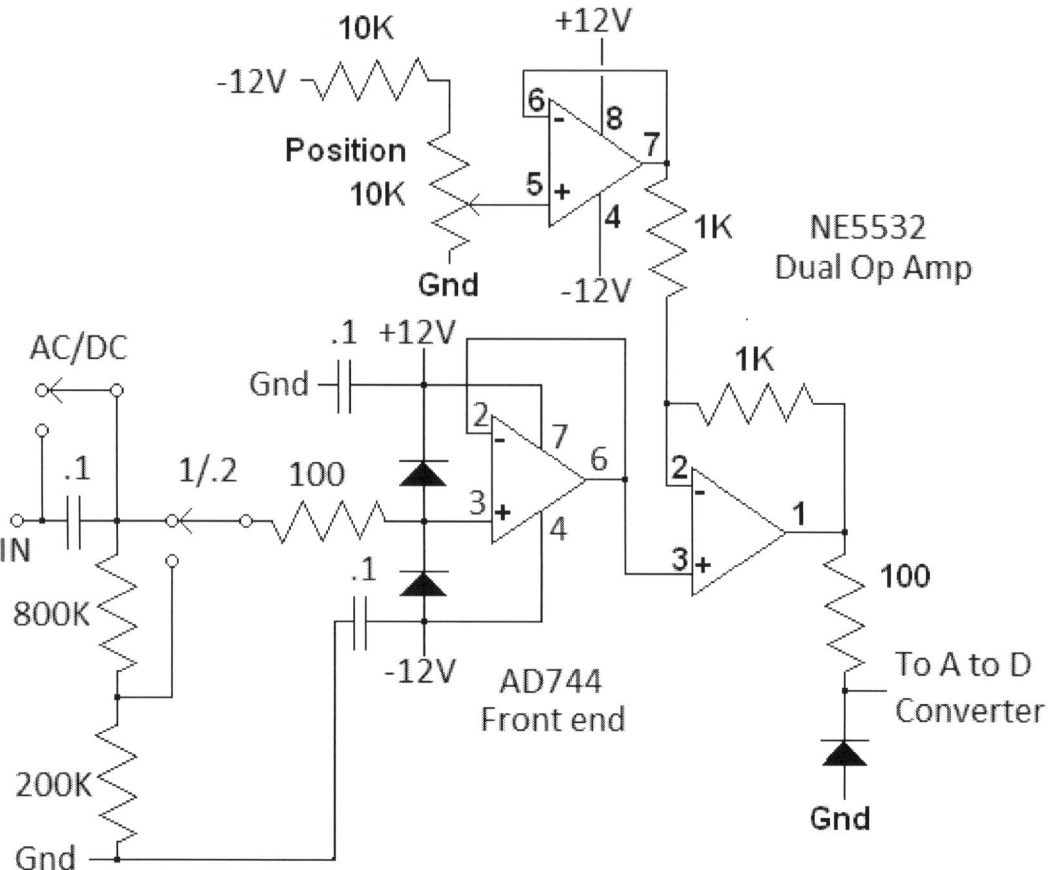

There is one more design option to consider. You can add a "gain selection" amplifier in between the AD744 and the NE5532. This optional amplifier will give you selectable gain of .5, 1, or 2. Some oscilloscope designs have this stage set up to give gain selection of 1, 2, or 5. This extra amplifier makes it possible to see smaller signals.

However adding the extra amplifier will leave us short one OP amp. We will need to add back half of a TL082 for the bias buffer. The bias buffer op amp does not need to have a good frequency response because it is only being used to buffer the DC bias voltage.

Another problem with the gain amplifier is that it is an inverting amplifier. So then the bias summing amplifier needs to also be an inverting amplifier. That way the two inversions will cancel each other out. Up next is the schematic diagram showing the added gain selection amplifier and the inverted bias summing amplifier.

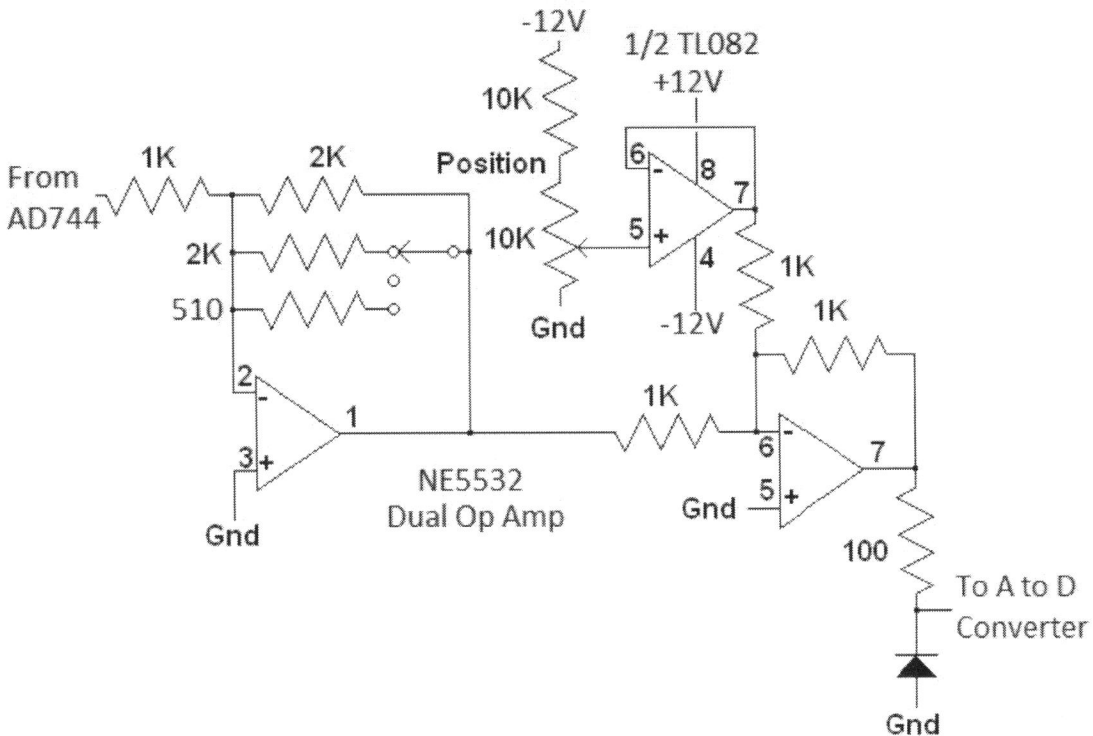

Here is a picture of the AD744 with NE5532 front end on a breadboard. This is my favorite analog input configuration.

Chapter 3

Fast Analog to Digital Converters

Here is a list of some of the fast analog to digital converters that I have tested out. There are others that I have tested, but some of them went up in smoke. The AD775 and TLC5510 are interchangeable. They are my favorites because they appear to take a lot of abuse and still keep on working.

CA3306 (Works best with port C "PINC" as it is 6 bits)
  Bits: 6
  Clock: 15MHz tested to 16MHz

CA3318 (8 bit version of the CA3306)
  Bits: 8
  Clock: 15MHz tested to 16MHz

AD775 (Out of production and hard to find)
  Bits: 8
  Clock: 20MHz tested to 25MHz

TDA8703 (Likes to keep on running without a clock)
  Bits: 8
  Clock: 40MHz tested to 50MHz

TLC5510 (Same pinout as CDX1175/ADC1175/ADC1173)
  Bits: 8
  Clock: 20MHz

TLC5540 (Same as LTC5510 but faster)
  Bits: 8
  Clock: 40MHz

ADC1175-50 (Same pinout as CDX1175/ADC1175/ADC1173)
  Bits: 8
  Clock: 50 MHz

Here is the schematic of the CA3306 analog to digital converter. The schematic is adapted from the PDF specifications file for the CA3306. The clock input can come from any 4 MHz to 15 MHz clock oscillator.

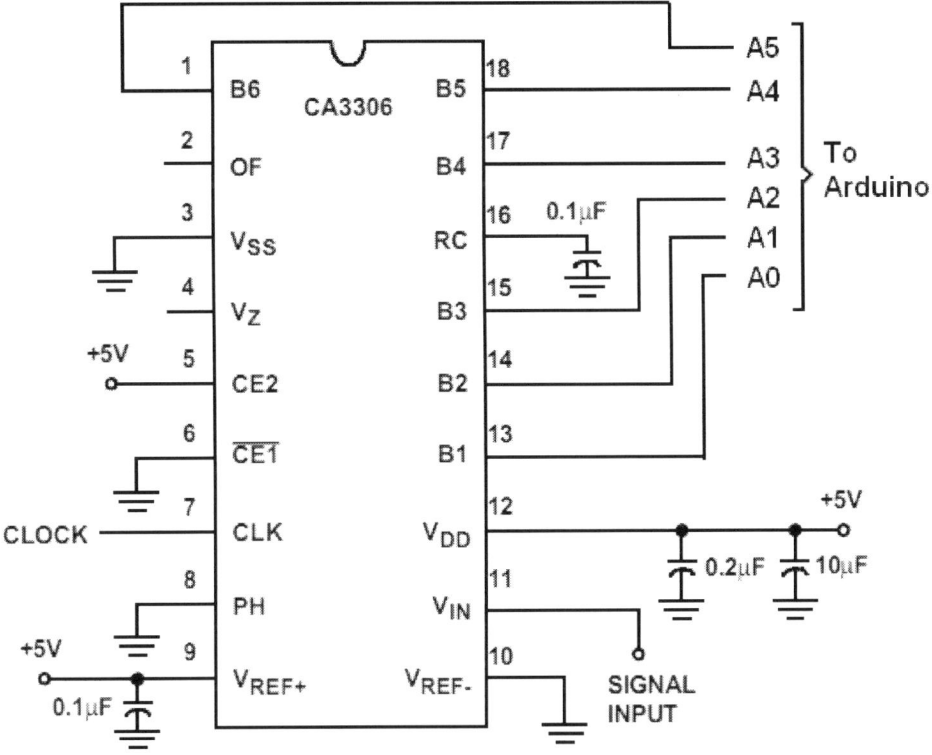

Here is a picture of the CA3306 analog to digital converter with an oscillator and filter capacitors.

The next Analog to digital converter is the CA3318. It is essentially a CA3306 but with 8 bits of accuracy. I have tested it at 16MHz but it fails at 20MHz. Here is the schematic diagram. The unmarked capacitors are .1 uF.

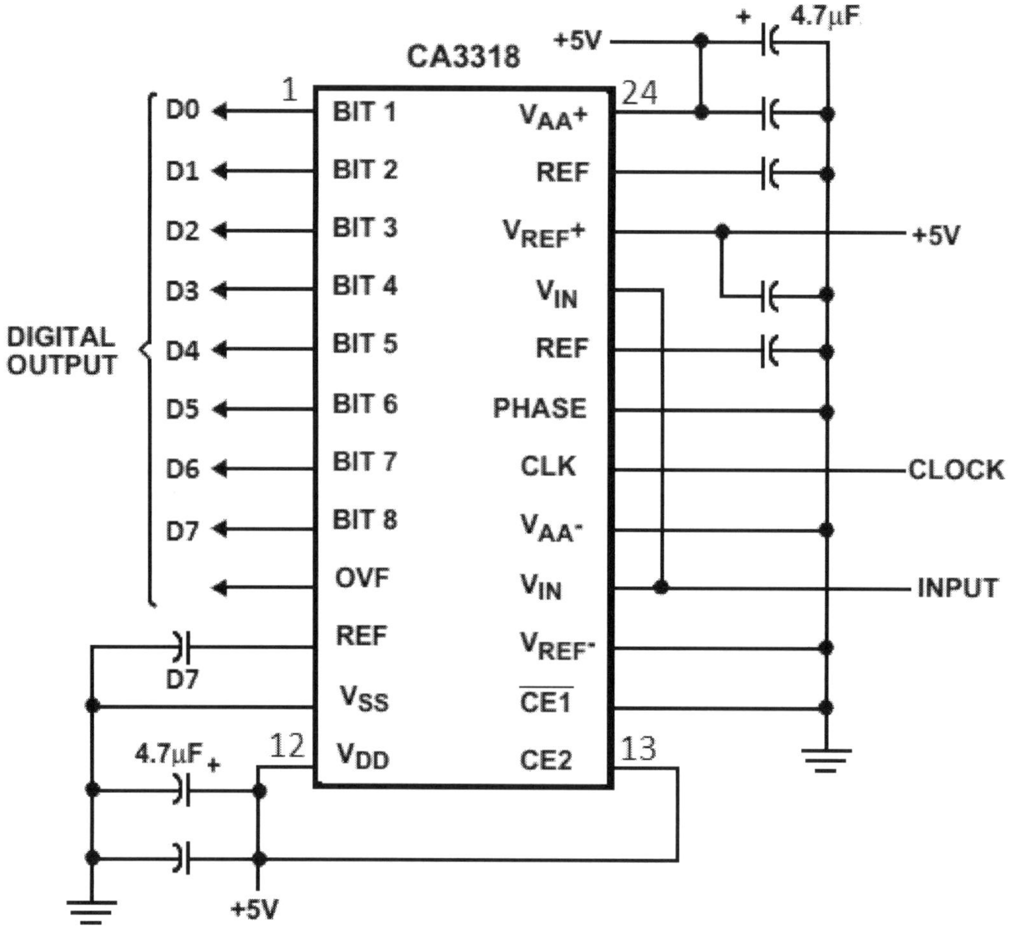

Here is a picture of the CA3318 wired up and working.

The AD775 was the first fast analog to digital converter that I experimented with. I made a parallel port video adapter using it way back in the 90's. Note that the pinout is almost identical to the TLC5510.

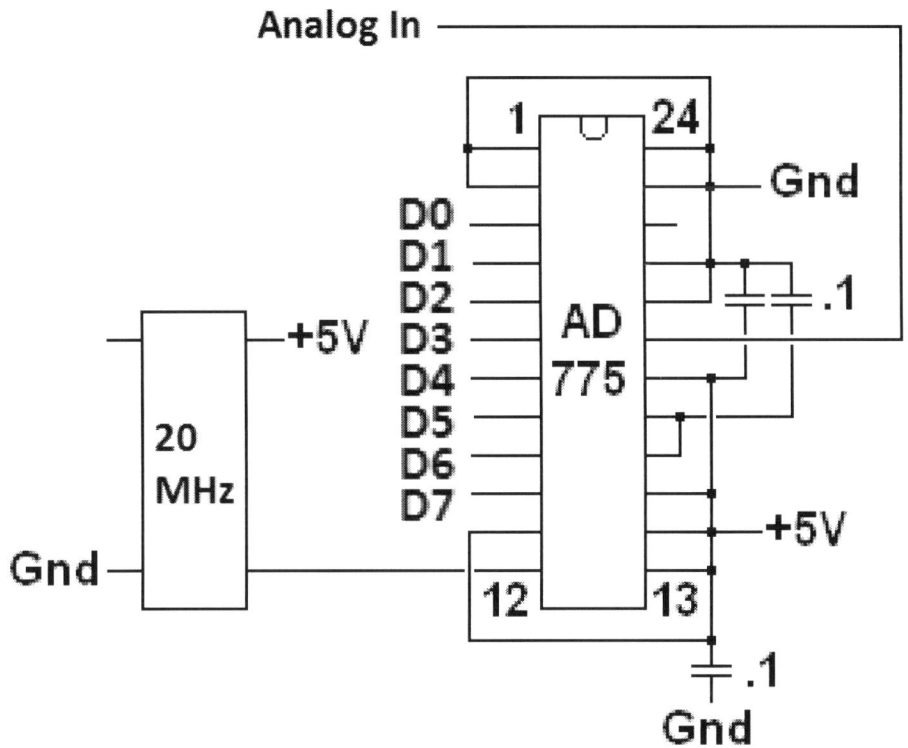

Here is a picture of the AD775 wired up and running.

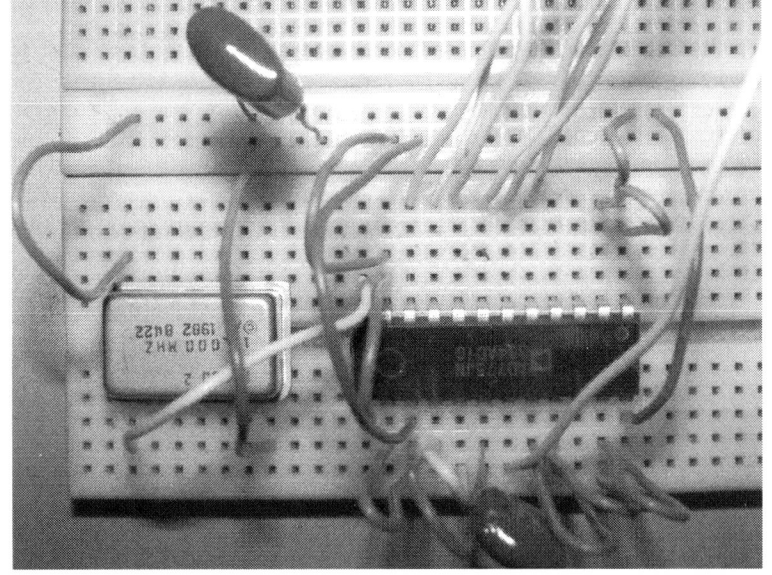

The TDA8703 is a physically larger IC. There is a small version available but it requires an adapter. The TDA8703 has been a very popular fast analog to digital converter for several years. It has been used in other digital oscilloscope designs. Here is the TDA8703 schematic diagram. "CK" is the Clock input.

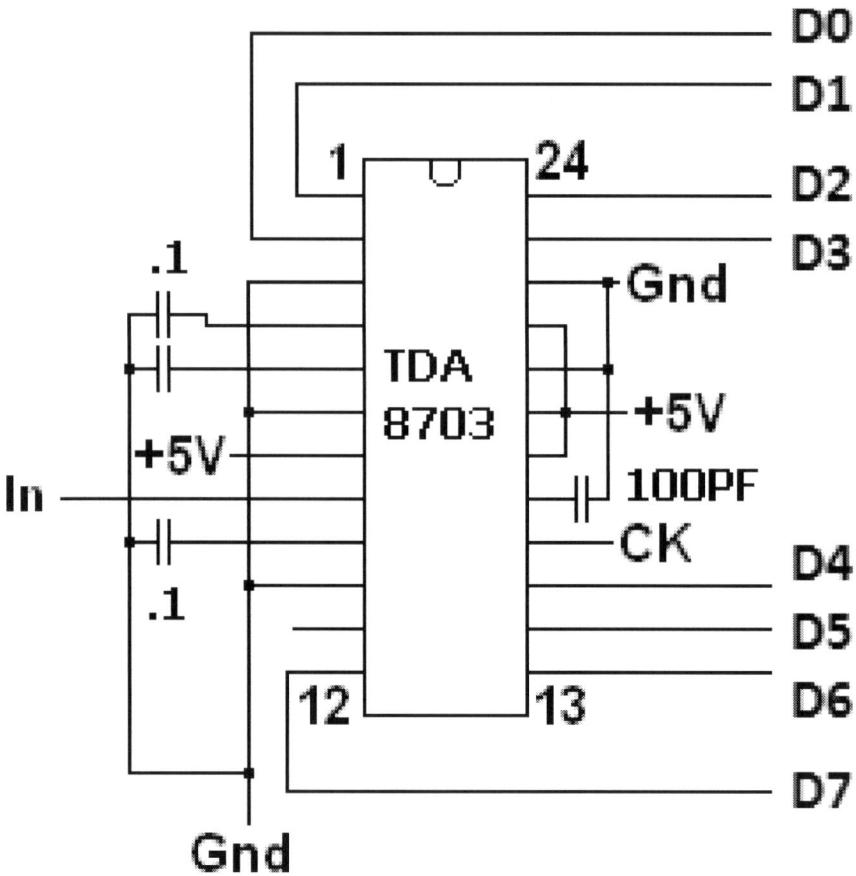

Up next is a picture of the TDA8703 wired to the analog input ports of an Arduino Uno. On the left side of the picture you can see the 10pf capacitor connected to the Arduino clock pin 9 to synchronize the two clocks.

The TLC5510 is usually a small outline IC (SOIC). You will need to use an adapter to plug it into a breadboard. Soldering the IC to the adapter can be a tricky operation. One secret is to use a liquid solder flux. Note that the schematic diagram shows the IC drawn upside down. I wired it up upside down to match the schematic and then spun the breadboard around to hook it up. The capacitors are all .1uf at 16 volts.

Also observe the addition of two 1K (or 2.2K) resistors, they help prevent D0 and D1 from messing with uploading software updates to the Arduino Uno. They are not needed if you disconnect D0 and D1 while uploading software. They do not solve the problem completely; sometimes you may still have to disconnect them to upload new software.

Up next is the TLC5510 Schematic based upon its application notes.

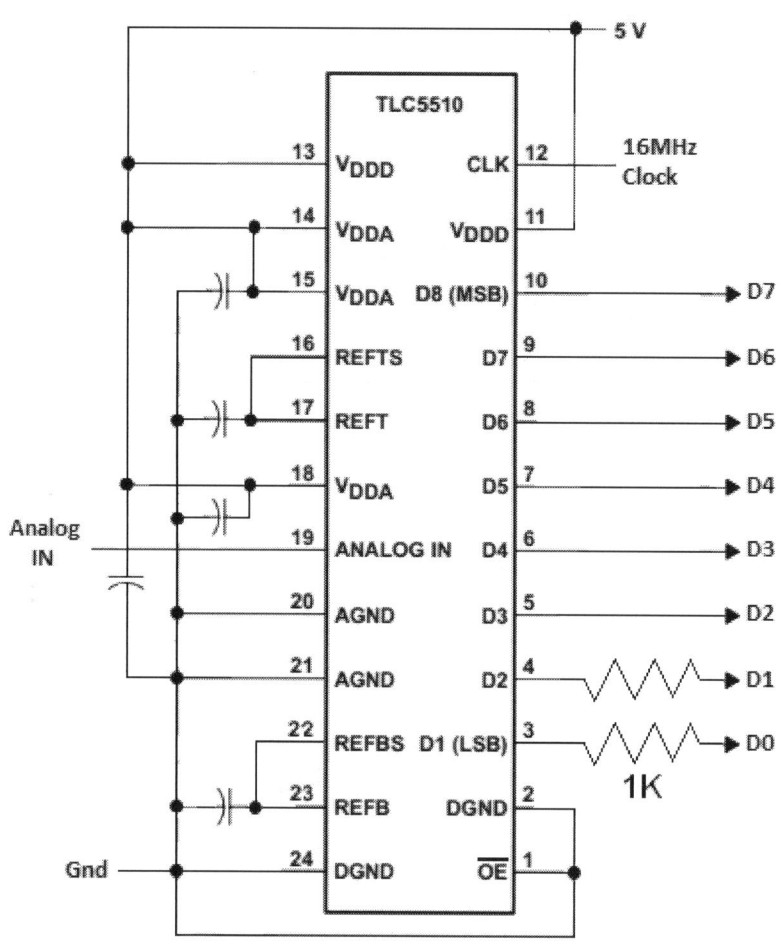

Up next is a picture of the TLC5510 wired up and running. In this picture you can also see the 10 pf capacitor going from the clock to pin 9 of the Arduino Uno to synchronize the clocks when not using a FIFO.

One of the problems with any free running analog to digital converters is "Clock Glitches". The glitch problem is caused by the clock for the analog to digital converter not properly matching the clock for the Arduino. Both the processor and the analog to digital converters are running in their "free run" mode.

However the processor will always read the contents of an input port on the same phase from its clock. The analog to digital converter will always update its results on the same phase from its clock. However if those two phases do not properly match an error will occur. If the analog converter output was in the process of changing from 10000000 to 01111111 the results that are read could be all 0's or all 1's during the time that it is updating. The solution would be to phase lock the two clocks and to make sure that the analog update phase does not match the Arduino parallel input phase.

For a long time I tried to tap into the Arduino clock on pin 9 and use that clock for running the analog to digital converter. Then I tried overriding the Arduino clock with an external clock but that did not work either. Then I found a simple solution to the problem. If you couple an external 16 MHz clock through a 10 or a 20 pf capacitor to pin 9 of the Arduino chip it works! The problem is that the clock input pin 9 is used to having a very small signal on it, so the external clock has to be very small as well. This solution only works for the Arduino Uno as it is a specific hardware modification.

Up next is a picture of what a clock glitch looks like on a LCD screen.

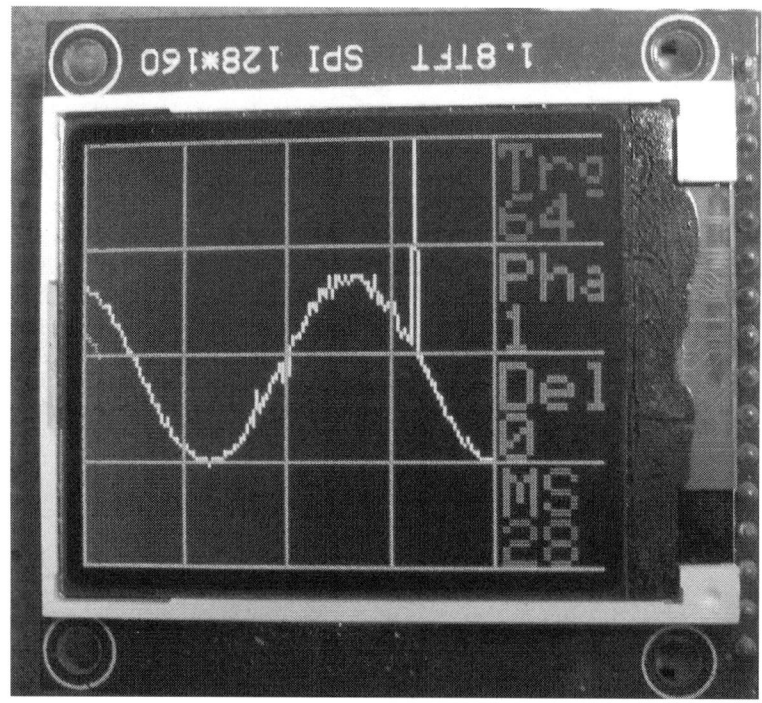

The improved clock synchronizing circuit is just a 16MHz oscillator with a 10 pf capacitor that goes to the Arduino IC clock pin 9. You could also use a 32 MHz clock and divide it by two first. This solution works with any of the analog to digital converters covered in this book.

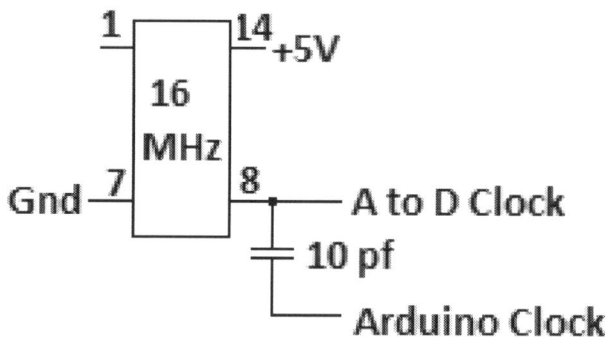

Chapter 4

Monochrome LCD QC12864B

Saying that graphics LCD's vary widely in their pinout is an understatement. When it comes to 128 by 64 graphics LCD's alone, they vary in their pinout, their physical size, the controller chip that is used, and even in their various modes of operation.

When I started out I did not have a 128 by 64 graphics display in stock, so I had to buy one for these projects. The first LCD arrived within a few weeks, but when I opened it, there was a 40 pin ribbon cable in the box instead of the LCD! I quickly went to one of my favorite vendors and ordered another 128 by 64 LCD without even reading the fine print. When it arrived I connected it up and loaded the software, but all I got was garbage on the screen. So I went back and read the fine print. The LCD was labeled QC12864B and it uses a ST7920 controller chip!

The solution is to use the universal Graphics library called "U8Glib.zip". That stands for Universal 8 bit graphics library. Once again, you need to decompress it into your "Arduino\libraries" directory and then make sure that it shows up in the Arduino interface.

Up next is a picture of that U8Glib library showing what it looks like when it is properly installed.

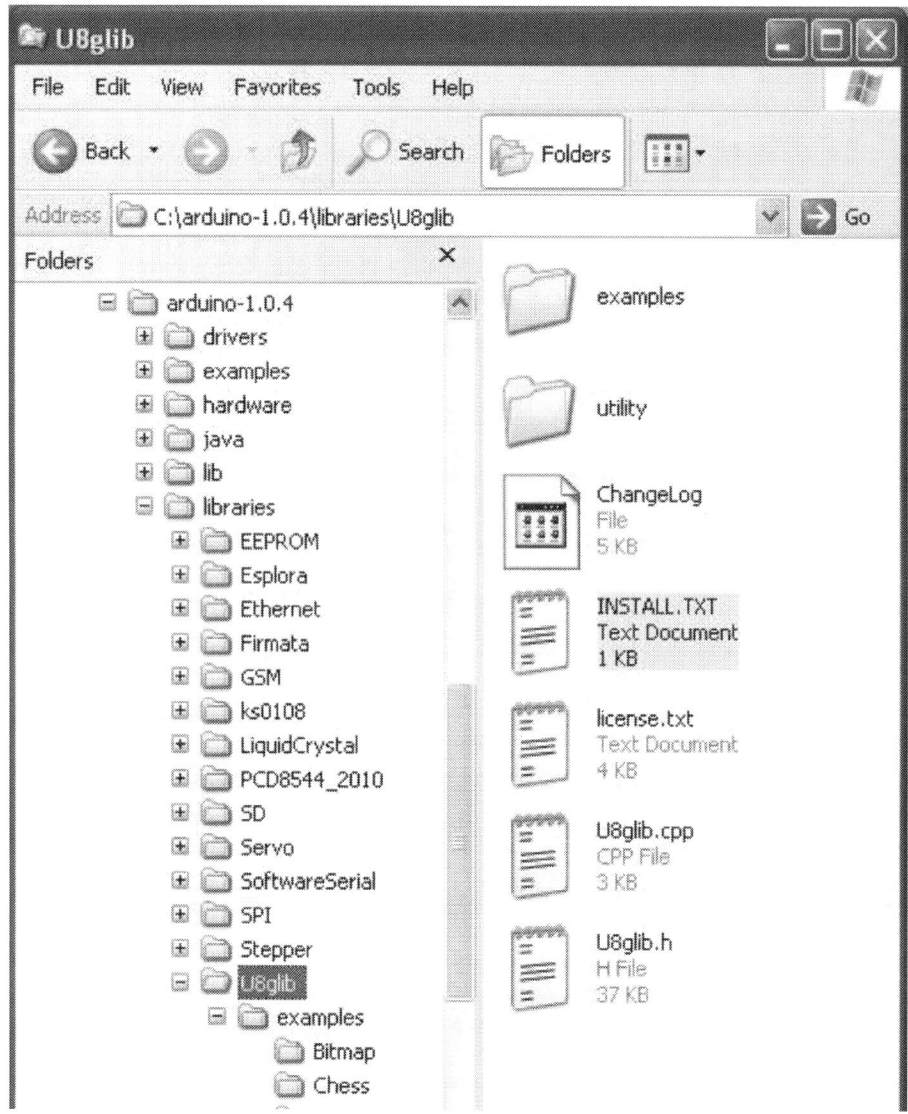

The U8G Library has a very long list of commands that it supports. Here is a quick list of the available commands:

Begin(void);
disableCursor(void);
drawBitmap, drawBitmapP(x, y, count, height, *bitmap);
drawBox(x, y, width, height);
drawCircle(x, y, radius);
drawDisc(x, y, radius);
drawFrame(x, y, width, height);
drawHLine, drawVLine(x, y, width);

drawLine(x1, y1, x2, y2);
drawPixel(x, y);
drawRBox, drawRFrame(x, y, width, height, radius);
drawStr, drawStr90, drawStr180, drawStr270(x, y, *string);
drawStrP, drawStr90P, drawStr180P, drawStr270P(x, y, *string);
drawXBM, drawXBMP(x, y, width, height, *bitmap);
enableCursor(void);   firstPage(void);
getColorIndex(void);   getFontAscent(void);
getFontDescent(void);   getFontLineSpacing(void);
getHeight(void);   getMode(void);
getWidth(void);
getStrWidth(*string);
InitSPI(*dev, sck, mosi, cs, a0, reset);
InitHWSPI(*dev, cs, a0, reset);
Init8Bit(*dev, d1, d2, d3, d4, d5, d6, d7, en, cs1, cs2, di, rw, reset);
nextPage(void);
print(. . .);
setColorIndex(Color_index);
setContrast(contrast);
setCursorColor(foreground,background);
setCursorFont(*font);
setCursorPos(x, y);
setCursorStyle(encoding);
setDefaultBackgroundColor, setDefaultForegroundColor(void);
setDefaultMidColor(void);
setFont(*font);
setFontLineSpacingFactor(factor)
setFontPosBaseline, setFontPosBottom(void);
setFontPosCenter, setFontPosTop(void);
setFontRefHeightAll, setFontRefHeightExtendedText(void);
setFontRefHeightText(void);
setPrintPos(x, y);
setRot90, setRot180, setRot270();
setScale2x2();
sleepOn, sleepOff(void);
undoRotation();
undoScale();

There is also a long list of driver chips and graphics formats that U8Glib supports. You have to remove the "//" in front of the chip and graphics setup that you are using. Then you wire up the LCD according to the pinout that is given on the next line. This is only a partial list of the many supported chips. We will be using the ST7920 chipset so it does not have the // in front if it.

U8GLIB_ST7920_128X64_4X u8g(8, 9, 10, 11, 4, 5, 6, 7, 18, 17, 16);
// 8Bit Com: D0..D7: 8,9,10,11,4,5,6,7 en=18, di=17,rw=16
//U8GLIB_PCD8544 u8g(13, 11, 10, 9, 8);
// SPI Com: SCK = 13, MOSI = 11, CS = 10, A0 = 9, Reset = 8
//U8GLIB_KS0108_128 u8g(8, 9, 10, 11, 4, 5, 6, 7, 18, 14, 15, 17, 16);
// 8Bit Com: D0-D7:8,9,10,11,4,5,6,7 en=18,cs1=14,cs2=15,di=17,rw=16
//U8GLIB_LC7981_160X80 u8g(8, 9,10,11,4,5,6,7, 18, 14, 15, 17, 16);
//8Bit Com:D0..D7:8,9,10,11,4,5,6,7 en=18, cs=14,di=15,rw=17,reset=16
//U8GLIB_SSD1306_128X64 u8g(13, 11, 10, 9);
// SW SPI Com: SCK = 13, MOSI = 11, CS = 10, A0 = 9
//U8GLIB_SSD1309_128X64 u8g(13, 11, 10, 9);
// SPI Com: SCK = 13, MOSI = 11, CS = 10, A0 = 9
//U8GLIB_NHD_C12864 u8g(13, 11, 10, 9, 8);
// SPI Com: SCK = 13, MOSI = 11, CS = 10, A0 = 9, RST = 8
//U8GLIB_NHD_C12832 u8g(13, 11, 10, 9, 8);
// SPI Com: SCK = 13, MOSI = 11, CS = 10, A0 = 9, RST = 8

Below is a picture of the pinout sheet for the QC12864B 128 by 64 graphics LCD.

ITEM	SYMBOL	LEVEL	FUNCTIONS
1	VSS	0V	Power Ground
2	VDD	+5V	Power supply for logic
3	V0	—	Contrast adjust
4	RS(CS)	H/L	H:data L:command
5	RW/(SID)	H/L	H:read L:write
6	E/(SCLK)	H.H→L	Enable signal
7-14	DB0-DB7	H/L	Data Bus
15	PSB	H/L	H:Paraller mode L:serial mode
16	NC	—	No connection
17	/REST	L	Reset signal
18	VOUT	—	Output LCD voltage
19	LEDA	+5V	Power supply for LED backlight
20	LEDK	0V	

If you think that chart is confusing, then you are not alone. On top of that, we have the connections to the Arduino UNO to consider. Here is the pinout chart showing the Arduino connections written in plain English.

Pin	Symbol	Arduino	Function
1	Vss	GND	Ground
2	Vdd	5V	Five Volts
3	Vo		Contrast variable resistor wiper
4	RS/CS	17 - A3	Register Select High is data, low is a command
5	RW/SID	16 - A2	Read Write, or Serial Data High is read, Low is write
6	E/SCLK	18 - A4	Enable or Serial Clock
7	DB0	8	data Bus bit 0
8	DB1	9	data Bus bit 1
9	DB2	10	data Bus bit 2
10	DB3	11	data Bus bit 3
11	DB4	4	data Bus bit 4
12	DB5	5	data Bus bit 5
13	DB6	6	data Bus bit 6
14	DB7	7	data Bus bit 7
15	PSB	5V	Parallel/Serial Bus High is Parallel, Low is Serial
16	NC		Not Connected – a really easy one
17	RST	5V	Reset Signal – Low is Reset
18	Vout		Voltage out to the contrast variable resistor
19	LED+	5V	Anode to the LED backlight
20	LED	GND	Ground for the LED backlight

Once the support software is installed you can wire it up and plug it in. My LCD still produced garbage, but it was because I had some data wires swapped. I used all white jumpers. Some people recommend using different color jumper wires for each bit so they do not get crossed by accident.

Up next is the schematic diagram for the QC12864B in parallel eight bit operation.

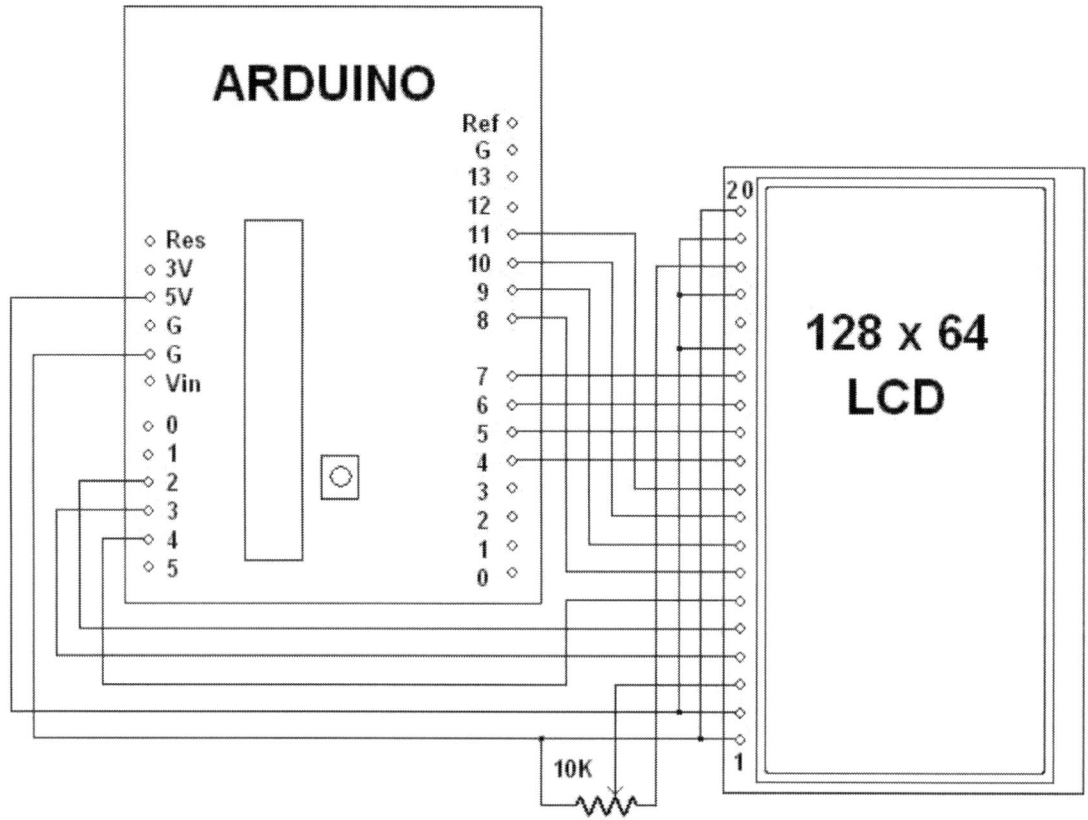

This next project is a simple Arduino based oscilloscope. It has a software based trigger loop so the waveform does not jump around nearly as much.

To test out this project I was running a program called "Sweep Generator" on my laptop computer. The program name is "sweepgen.exe". I have tested this oscilloscope design with an audio sweep range varying from 100 Hz to 1000 Hz. This simple oscilloscope is quite useable up to about 1 KHz, but it is not very usable above that frequency. The display starts looking like a lot of noise above 1 KHz.

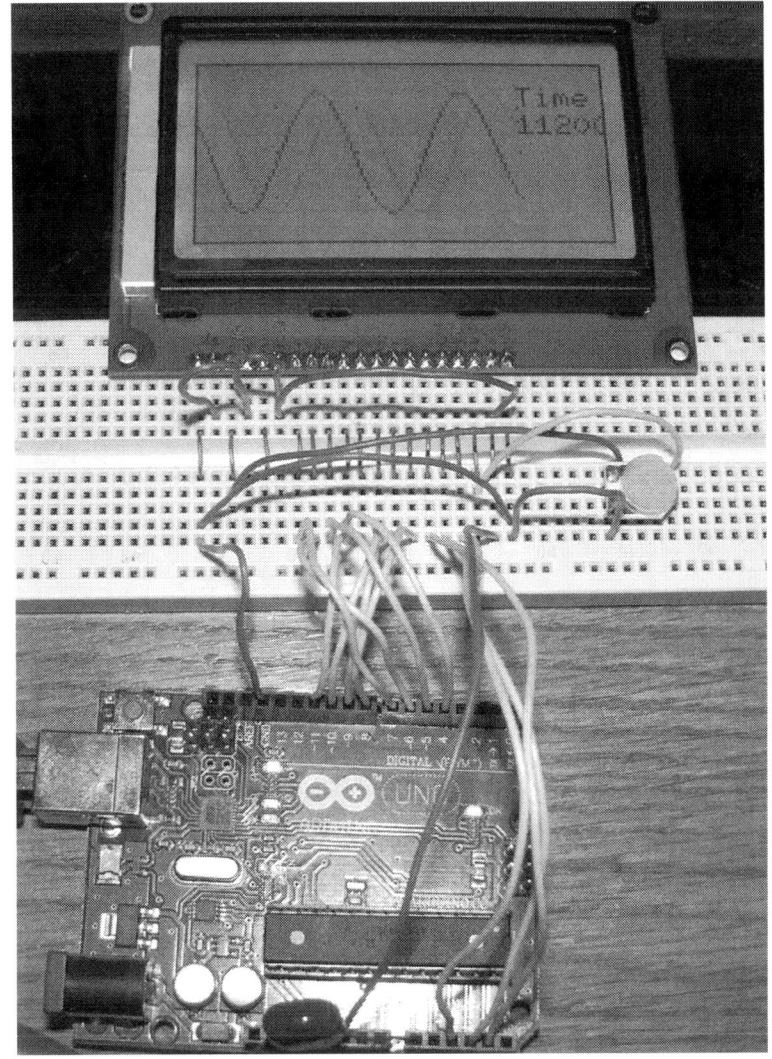

Here is the simple oscilloscope sketch.

/*********************************
128 by 64 LCD Oscilloscope - Default
By Bob Davis
Uses Universal 8bit Graphics Library, http://code.google.com/p/u8glib/
  Copyright (c) 2012, olikraus@gmail.com   All rights reserved.

*********************************/
#include "U8glib.h"

// 8Bit Com: D0..D7: 8,9,10,11,4,5,6,7 en=18, di=17,rw=16

```
U8GLIB_ST7920_128X64_4X u8g(8, 9, 10, 11, 4, 5, 6, 7, 18, 17, 16);
int Sample[128];
int Input=0;
int OldInput=0;
int StartSample=0;
int EndSample=0;
int SampleTime=0;
void u8g_prepare(void) {
  u8g.setFont(u8g_font_6x10);
  u8g.setFontRefHeightExtendedText();
  u8g.setDefaultForegroundColor();
  u8g.setFontPosTop();
}
void DrawMarkers(void) {
  u8g.drawFrame (0,0,128,64);
  u8g.drawPixel (25,16);
  u8g.drawPixel (50,16);
  u8g.drawPixel (100,16);
  u8g.drawPixel (25,32);
  u8g.drawPixel (50,32);
  u8g.drawPixel (100,32);
  u8g.drawPixel (25,48);
  u8g.drawPixel (50,48);
  u8g.drawPixel (100,48);
}
void sample_data(void){
// wait for a trigger of a positive going input
  while (Input < OldInput){
    OldInput=analogRead(A0);
    Input=analogRead(A0);
  }
// collect the analog data into an array
// do not do division by 10.24 here, it makes it slower!
  StartSample = micros();
  for(int xpos=0; xpos<100; xpos++) {
    Sample[xpos]=analogRead(A0);
  }
  EndSample = micros();
}
void draw(void) {
  char buf[12];
  u8g_prepare();
```

```
  DrawMarkers();
// display the collected analog data from array
// Sample/10.24 because 1024 becomes 100 = 5 volts
  for(int xpos=1; xpos<99; xpos++) {
    u8g.drawLine (xpos, Sample[xpos]/10.24, xpos+1, Sample[xpos+1]/10.24);
  }
  SampleTime=EndSample-StartSample;
  u8g.drawStr(100, 8, "Time");
  u8g.drawStr(100, 18, itoa(SampleTime, buf, 10));
}
void setup(void) {
  // assign default color value
  if ( u8g.getMode() == U8G_MODE_R3G3B2 )
    u8g.setColorIndex(255);     // RGB=white
  else if ( u8g.getMode() == U8G_MODE_GRAY2BIT )
    u8g.setColorIndex(3);       // max intensity
  else if ( u8g.getMode() == U8G_MODE_BW )
    u8g.setColorIndex(1);       // pixel on, black
}
void loop(void) {
// collect the data
  sample_data();
// show collected data
  u8g.firstPage();
  do { draw(); }
  while( u8g.nextPage() );
// rebuild the picture after some delay
  delay(100);
}
```

The next project is a six channel logic analyzer. We need to make all six of the analog inputs available so we will need to move the control signals for the LCD to other pins. D1, D2 and D3 are all available with this LCD so I moved the control signals there. Here is an updated schematic with the control pins moved. This change leaves all six analog inputs free to be used as logic inputs.

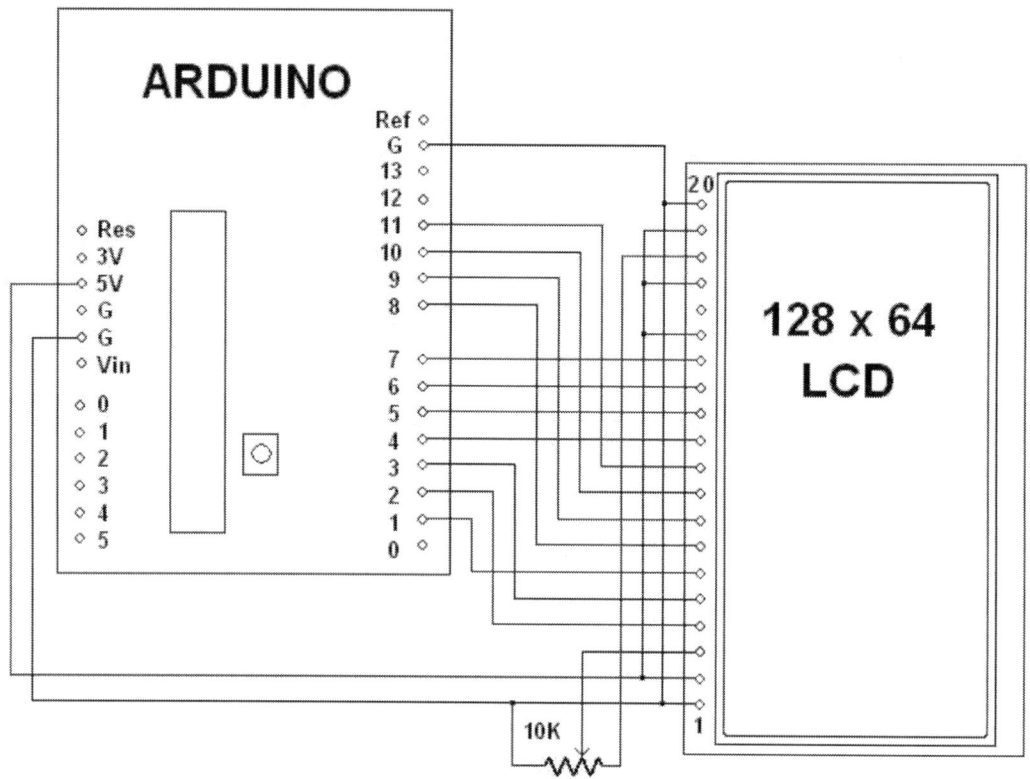

Coming up next is a picture of the six channel logic analyzer in use. The picture was taken while it was monitoring the 100 KC to 1 MC outputs of a 74LS390 dual divide by 10 IC with a 10 Mc Clock input. The resulting display is quite interesting.

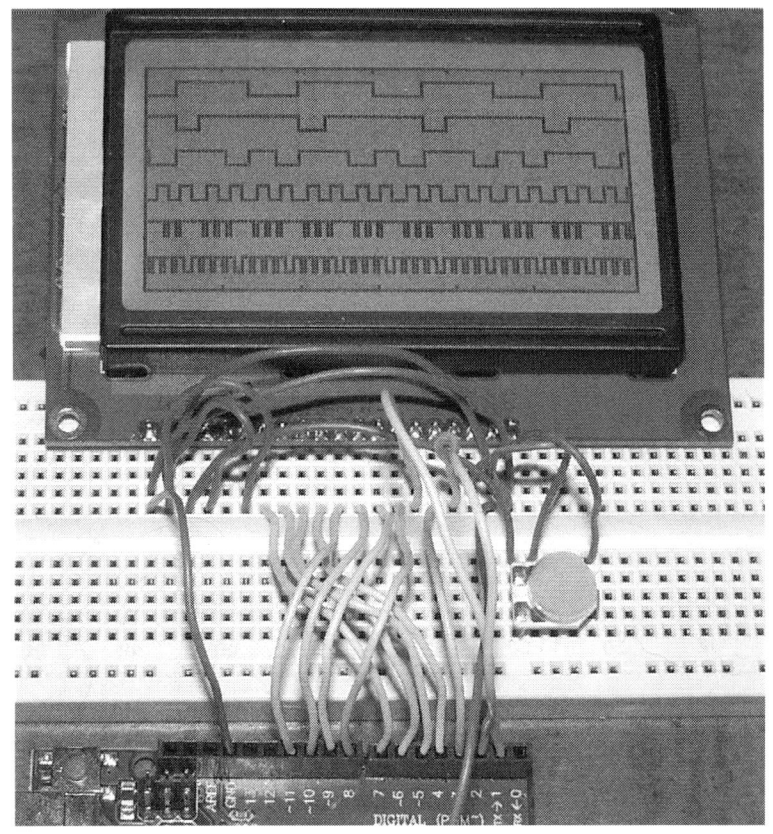

Here is the sketch code to make the six channel logic analyzer work.

```
/***********************************
128 by 64 six channel LCD Logic Analyzer
By Bob Davis
Uses Universal 8bit Graphics Library, http://code.google.com/p/u8glib/
  Copyright (c) 2012, olikraus@gmail.com    All rights reserved.
*********************************************/
#include "U8glib.h"

// 8Bit Com: D0..D7: 8,9,10,11,4,5,6,7 en=18, di=17,rw=16
//U8GLIB_ST7920_128X64_4X u8g(8, 9, 10, 11, 4, 5, 6, 7, 18, 17, 16);
//  **** NOTE **** I Moved the three control pins !!!
U8GLIB_ST7920_128X64_4X u8g(8, 9, 10, 11, 4, 5, 6, 7, 1, 2, 3);

int Sample[128];
int Input=0;
int OldInput=0;
```

```
void setup(void) {
  pinMode(A0, INPUT);
  pinMode(A1, INPUT);
  pinMode(A2, INPUT);
  pinMode(A3, INPUT);
  pinMode(A4, INPUT);
  pinMode(A5, INPUT);
  // assign the default color value
  if ( u8g.getMode() == U8G_MODE_R3G3B2 )
    u8g.setColorIndex(255);     // RGB=white
  else if ( u8g.getMode() == U8G_MODE_GRAY2BIT )
    u8g.setColorIndex(3);       // max intensity
  else if ( u8g.getMode() == U8G_MODE_BW )
    u8g.setColorIndex(1);       // pixel on, black
}

void u8g_prepare(void) {
  u8g.setFont(u8g_font_6x10);
  u8g.setFontRefHeightExtendedText();
  u8g.setDefaultForegroundColor();
  u8g.setFontPosTop();
}

void DrawMarkers(void) {
  u8g.drawFrame (0,0,128,64);
  u8g.drawPixel (20,1);
  u8g.drawPixel (40,1);
  u8g.drawPixel (60,1);
  u8g.drawPixel (80,1);
  u8g.drawPixel (100,1);
  u8g.drawPixel (20,62);
  u8g.drawPixel (40,62);
  u8g.drawPixel (60,62);
  u8g.drawPixel (80,62);
  u8g.drawPixel (100,62);
}

void draw(void) {
  u8g_prepare();
  DrawMarkers();
  // wait for a trigger of a positive going input
  Input=digitalRead(A0);
```

```
  while (Input != 1){
    Input=digitalRead(A0);
  }

  // collect the analog data into an array
  for(int xpos=0; xpos<128; xpos++) {
    Sample[xpos]=PINC;
  }

  // display the collected analog data from the array
  for(int xpos=0; xpos<128; xpos++) {
    u8g.drawLine (xpos, ((Sample[xpos]&B00000001)*4)+4, xpos+1, ((Sample[xpos+1]&B00000001)*4)+4);
    u8g.drawLine (xpos, ((Sample[xpos]&B00000010)*2)+14, xpos+1, ((Sample[xpos+1]&B00000010)*2)+14);
    u8g.drawLine (xpos, ((Sample[xpos]&B00000100)*1)+24, xpos+1, ((Sample[xpos+1]&B00000100)*1)+24);
    u8g.drawLine (xpos, ((Sample[xpos]&B00001000)/2)+34, xpos+1, ((Sample[xpos+1]&B00001000)/2)+34);
    u8g.drawLine (xpos, ((Sample[xpos]&B00010000)/4)+44, xpos+1, ((Sample[xpos+1]&B00010000)/4)+44);
    u8g.drawLine (xpos, ((Sample[xpos]&B00100000)/8)+54, xpos+1, ((Sample[xpos+1]&B00100000)/8)+54);
  }
} // end of draw routine

void loop(void) {
// this is the picture loop
// u8g.firstPage();
  do { draw(); }
  while( u8g.nextPage() );
  // rebuild the picture after some delay
  delay(100);
}
```

You can take the six channel logic analyzer a step further and add an external CA3306 analog to digital converter. The CA3306 schematic is in the third chapter of this book. You can also add speed select switches on D12 and D13 to ground. Here is a picture of the LCD screen running the 5 MSPS oscilloscope.

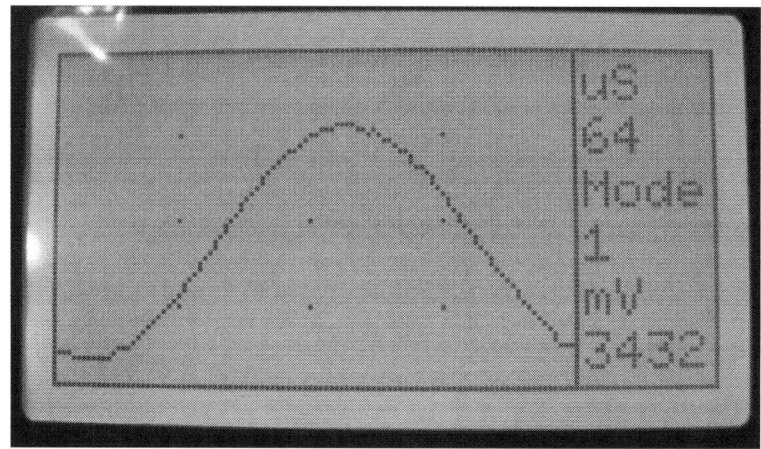

You can also use this sketch code for a five million sample per second logic analyzer or for an oscilloscope.

```
/*********************************
128 by 64 LCD Oscilloscope ext AtoD speed select
By Bob Davis
Uses Universal 8bit Graphics Library, http://code.google.com/p/u8glib/
  Copyright (c) 2012, olikraus@gmail.com   All rights reserved.
*********************************/
#include "U8glib.h"

// 8Bit Com: D0..D7: 8,9,10,11,4,5,6,7 en=18, di=17,rw=16
//U8GLIB_ST7920_128X64_4X u8g(8, 9, 10, 11, 4, 5, 6, 7, 18, 17, 16);
// NOTE taht the pins have bee rearranged
U8GLIB_ST7920_128X64_4X u8g(8, 9, 10, 11, 4, 5, 6, 7, 1, 2, 3);

byte Sample[100];
//int Sample[100];
int Input=0;
int OldInput=0;
long StartSample=0;
long EndSample=0;
long SampleTime=0;
int MaxSample=0;
int MinSample=0;
int SampleSize=0;
int dTime=0;
int mode=0;
```

```
void u8g_prepare(void) {
  u8g.setFont(u8g_font_6x10);
  u8g.setFontRefHeightExtendedText();
  u8g.setDefaultForegroundColor();
  u8g.setFontPosTop();
}

void DrawMarkers(void) {
  u8g.drawFrame (0,0,128,64);
  u8g.drawFrame (100,0,128,64);
  u8g.drawPixel (25,16);
  u8g.drawPixel (50,16);
  u8g.drawPixel (75,16);
  u8g.drawPixel (25,32);
  u8g.drawPixel (50,32);
  u8g.drawPixel (75,32);
  u8g.drawPixel (25,48);
  u8g.drawPixel (50,48);
  u8g.drawPixel (75,48);
}

void get_mode(void) {
   if (digitalRead(13) == 0) mode++;
   if (digitalRead(12) == 0) mode--;
   if (mode > 9) mode = 0;
   if (mode < 0) mode = 9;
// Select delay times for loop modes
   if (mode == 0) dTime=0;
   if (mode == 1) dTime=0;
   if (mode == 2) dTime=1;
   if (mode == 3) dTime=5;
   if (mode == 4) dTime=10;
   if (mode == 5) dTime=50;
   if (mode == 6) dTime=100;
   if (mode == 7) dTime=500;
   if (mode == 8) dTime=1000;
   if (mode == 9) dTime=5000;
}
void sample_data(void){
// wait for a trigger of a positive going input
   while (digitalRead(A0)==0) { }
// collect the analog data into an array
```

```
// mode 0 will use verbose method
if (mode == 0) {
  StartSample = micros();
    Sample[0]=PINC;
    Sample[1]=PINC;    Sample[2]=PINC;    Sample[3]=PINC;
    Sample[4]=PINC;    Sample[5]=PINC;    Sample[6]=PINC;
    Sample[7]=PINC;    Sample[8]=PINC;    Sample[9]=PINC;
    Sample[10]=PINC;   Sample[11]=PINC;   Sample[12]=PINC;
    Sample[13]=PINC;   Sample[14]=PINC;   Sample[15]=PINC;
    Sample[16]=PINC;   Sample[17]=PINC;   Sample[18]=PINC;
    Sample[19]=PINC;   Sample[20]=PINC;   Sample[21]=PINC;
    Sample[22]=PINC;   Sample[23]=PINC;   Sample[24]=PINC;
    Sample[25]=PINC;   Sample[26]=PINC;   Sample[27]=PINC;
    Sample[28]=PINC;   Sample[29]=PINC;   Sample[30]=PINC;
    Sample[31]=PINC;   Sample[32]=PINC;   Sample[33]=PINC;
    Sample[34]=PINC;   Sample[35]=PINC;   Sample[36]=PINC;
    Sample[37]=PINC;   Sample[38]=PINC;   Sample[39]=PINC;
    Sample[40]=PINC;   Sample[41]=PINC;   Sample[42]=PINC;
    Sample[43]=PINC;   Sample[44]=PINC;   Sample[45]=PINC;
    Sample[46]=PINC;   Sample[47]=PINC;   Sample[48]=PINC;
    Sample[49]=PINC;   Sample[50]=PINC;   Sample[51]=PINC;
    Sample[52]=PINC;   Sample[53]=PINC;   Sample[54]=PINC;
    Sample[55]=PINC;   Sample[56]=PINC;   Sample[57]=PINC;
    Sample[58]=PINC;   Sample[59]=PINC;   Sample[60]=PINC;
    Sample[61]=PINC;   Sample[62]=PINC;   Sample[63]=PINC;
    Sample[64]=PINC;   Sample[65]=PINC;   Sample[66]=PINC;
    Sample[67]=PINC;   Sample[68]=PINC;   Sample[69]=PINC;
    Sample[70]=PINC;   Sample[71]=PINC;   Sample[72]=PINC;
    Sample[73]=PINC;   Sample[74]=PINC;   Sample[75]=PINC;
    Sample[76]=PINC;   Sample[77]=PINC;   Sample[78]=PINC;
    Sample[79]=PINC;   Sample[80]=PINC;   Sample[81]=PINC;
    Sample[82]=PINC;   Sample[83]=PINC;   Sample[84]=PINC;
    Sample[85]=PINC;   Sample[86]=PINC;   Sample[87]=PINC;
    Sample[88]=PINC;   Sample[89]=PINC;   Sample[90]=PINC;
    Sample[91]=PINC;   Sample[92]=PINC;   Sample[93]=PINC;
    Sample[94]=PINC;   Sample[95]=PINC;   Sample[96]=PINC;
    Sample[97]=PINC;   Sample[98]=PINC;   Sample[99]=PINC;
//   Sample[100]=PINC;
  EndSample = micros();
}
// mode 1 will use loop with no delay
if (mode ==1) {
```

```
    StartSample = micros();
    for(int xpos=0; xpos<100; xpos++) {
      Sample[xpos]=PINC;
//    delayMicroseconds(dTime);
    }
    EndSample = micros();
  }
// mode 2 or more will use loop with delay
  if (mode >= 2) {
    StartSample = micros();
    for(int xpos=0; xpos<100; xpos++) {
      Sample[xpos]=PINC;
      delayMicroseconds(dTime);
    }
    EndSample = micros();
} }
void draw(void) {
  char buf[12];
  u8g_prepare();
  DrawMarkers();
// display the collected analog data from array
  for(int xpos=1; xpos<99; xpos++) {
//  For Oscope more use this line
    u8g.drawLine (xpos, Sample[xpos], xpos+1, Sample[xpos+1]);
//  For logic analyzer use the next 6 lines instead
//    u8g.drawLine (xpos, ((Sample[xpos]&B00000001)*4)+4, xpos, ((Sample[xpos+1]&B00000001)*4)+4);
//    u8g.drawLine (xpos, ((Sample[xpos]&B00000010)*2)+14, xpos, ((Sample[xpos+1]&B00000010)*2)+14);
//    u8g.drawLine (xpos, ((Sample[xpos]&B00000100)*1)+24, xpos, ((Sample[xpos+1]&B00000100)*1)+24);
//    u8g.drawLine (xpos, ((Sample[xpos]&B00001000)/2)+34, xpos, ((Sample[xpos+1]&B00001000)/2)+34);
//    u8g.drawLine (xpos, ((Sample[xpos]&B00010000)/4)+44, xpos, ((Sample[xpos+1]&B00010000)/4)+44);
//    u8g.drawLine (xpos, ((Sample[xpos]&B00100000)/8)+54, xpos, ((Sample[xpos+1]&B00100000)/8)+54);
  }
  SampleTime=EndSample-StartSample;
  if (SampleTime < 9999) u8g.drawStr(102, 2, "uS");
  if (SampleTime > 9999) {
    SampleTime=SampleTime/1000;
```

```
    u8g.drawStr(102, 2, "mS");
  }
  u8g.drawStr(102, 12, itoa(SampleTime, buf, 10));
  u8g.drawStr(102, 22, "Mode");
  u8g.drawStr(102, 32, itoa(mode, buf, 10));
// Determine sample voltage peak to peak
  MaxSample = Sample[10];
  MinSample = Sample[10];
  for(int xpos=0; xpos<100; xpos++) {
//    OldSample[xpos] = Sample[xpos];
    if (Sample[xpos] > MaxSample) MaxSample=Sample[xpos];
    if (Sample[xpos] < MinSample) MinSample=Sample[xpos];
    }
  // Range of 0 to 64 * 78 = 4992 mV
  SampleSize=(MaxSample-MinSample)*78;
  u8g.drawStr(102, 42, "mV");
  u8g.drawStr(102, 52, itoa(SampleSize, buf, 10));
}
void setup(void) {
  // set up input pins
  pinMode(12, INPUT);
  digitalWrite(12, HIGH);
  pinMode(13, INPUT);
  digitalWrite(13, HIGH);
  // assign default color value
  if ( u8g.getMode() == U8G_MODE_R3G3B2 )
    u8g.setColorIndex(255);     // RGB=white
  else if ( u8g.getMode() == U8G_MODE_GRAY2BIT )
    u8g.setColorIndex(3);       // max intensity
  else if ( u8g.getMode() == U8G_MODE_BW )
    u8g.setColorIndex(1);       // pixel on, black
}
void loop(void) {
// Set up the mode
  get_mode();
// collect the data
  sample_data();
// show collected data
  u8g.firstPage();
  do { draw(); }
  while( u8g.nextPage() );
// rebuild the picture after some delay
```

delay(500);
}
// end of program

Chapter 5

Serial Color LCD 1.8TFT SPI

This is a serial LCD screen so it will use less of the Arduino I/O pins leaving more pins free for other things. This LCD only needs five I/O pins along with power and ground in order to operate. The programs for this LCD will use the new built in TFT drivers that are found in version 1.0.5 and above of the Arduino driver. There appears to be two versions of the 1.8 inch TFT LCD screen. One version has 10 pins and the other version has 16 pins. The pin definitions are written on the bottom of the circuit board so it is just a matter of writing them down before you flip it over and wire it up.

Here is a picture of the back of the 1.8 inch LCD. As you can see the pins are clearly marked.

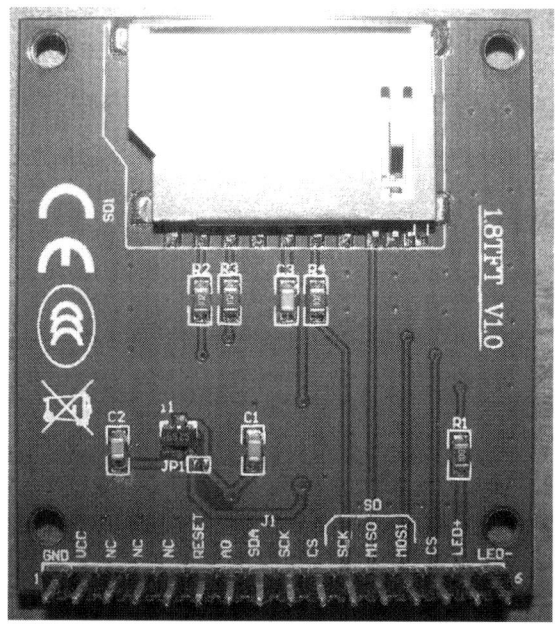

The sketches for this LCD use the new built in TFT drivers found in version 1.0.5 of the Arduino driver. The LCD screen will not work without this TFT driver being properly installed. This built in TFT driver is the same one as the Adafruit ST7735 driver, it is just renamed as "TFT".

Here is a chart showing the wiring from the Arduino Uno to the LCD screen. I ran jumpers to connect the two ground pins and 5V pins together on the breadboard.

```
Arduino Uno         1.8 SPI TFT
----------------    ---------------
GND                 Pin 01 (GND)
5V (VCC)            Pin 02 (VCC)
Not used            Pin 03
Not used            Pin 04
Not used            Pin 05
D8                  Pin 06 (RESET)
D9                  Pin 07 (A0)
D11 (MOSI)          Pin 08 (SDA)
D13 (SCK)           Pin 09 (SCK)
D10 (SS)            Pin 10 (CS)
Not used            Pin 11 SD Card
Not used            Pin 12 SD Card
Not used            Pin 13 SD Card
Not used            Pin 14 SD Card
5V (VCC)            Pin 15 (LED+)
GND                 Pin 16 (LED-)
```

Note that pin one of the LCD is located on the right side as you look at the top of the LCD screen.

Up next is the schematic diagram of how to wire this LCD screen up. Note that D0-D7 and the six analog pins A0-A6 are now all free to be used for this and for other projects.

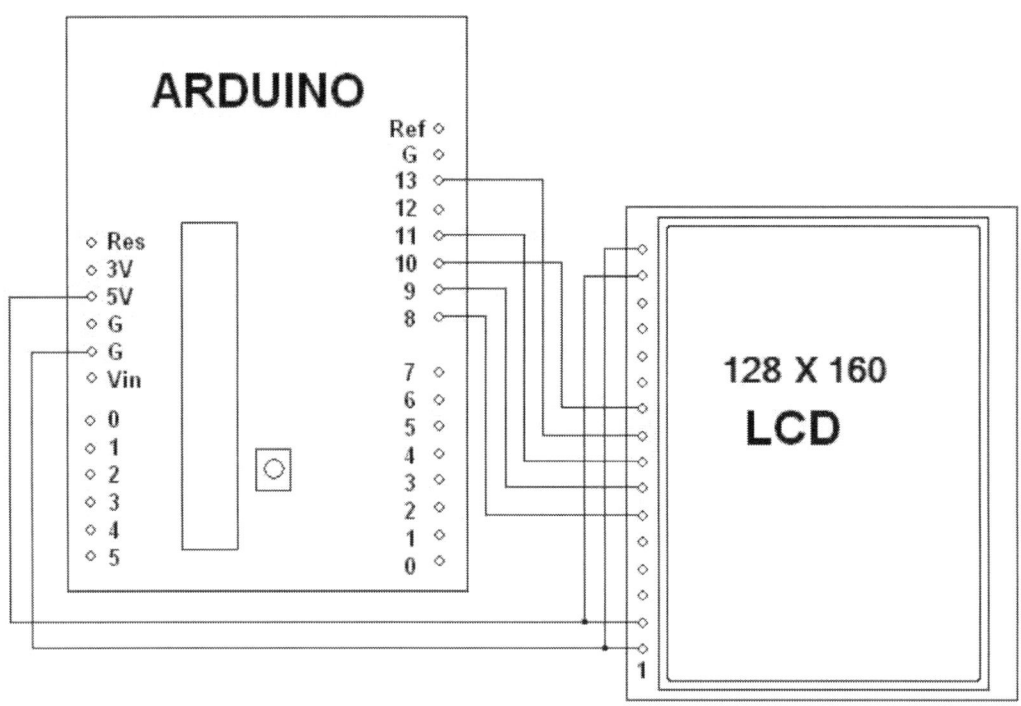

Up next is a picture of this LCD screen while running the simple oscilloscope demonstration sketch. This is a fairly simple oscilloscope using A0 as the analog input. For an analog input circuit see the analog input chapter earlier in this book.

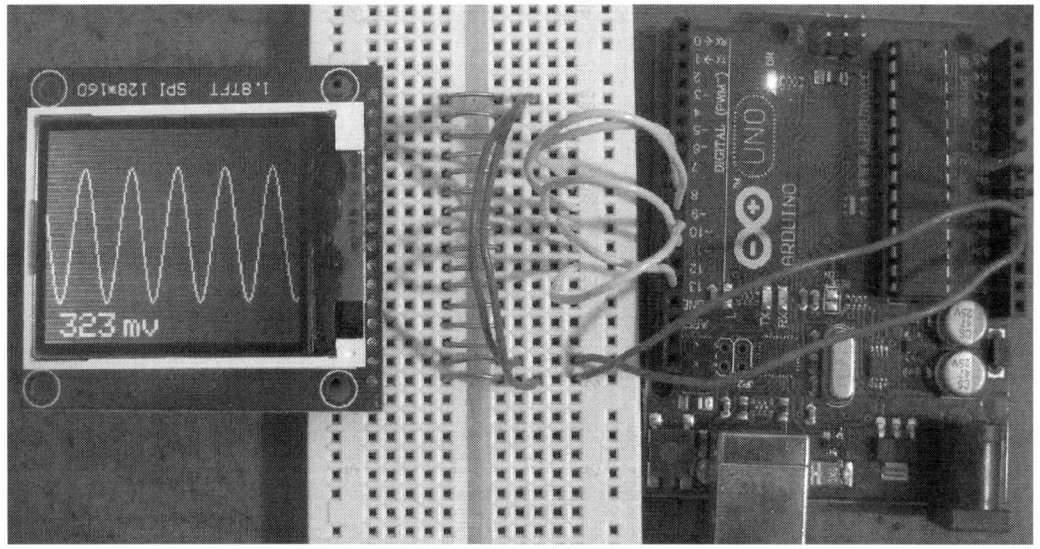

This sketch produces a simple oscilloscope that is good to about 1KHz. The trace is in red and the text listing the peak to peak voltage is in green.

```
/*********************************
TFT Oscilloscope
Reads the value of analog input on A0,
and shows the value on the screen.
Created 15 January 2014 by Bob Davis
**************************************/

#include <TFT.h>  // Arduino LCD library
#include <SPI.h>
 // pin definition for the Uno
#define rst  8
#define dc   9
#define cs   10
TFT TFTscreen = TFT(cs, dc, rst);

// set up variables
int xPos = 0;
int value = 0;
int maxvalue=100;
int minvalue=100;
int sensor[160];
char buf[12];

void setup(){
  // initialize the display
  TFTscreen.begin();
  // clear the screen
  TFTscreen.background(0, 0, 0);
  // Set the font size
  TFTscreen.setTextSize(2);
}

void loop(){
  // quickly collect the data
  for (int xpos = 0; xpos <160; xpos++){
  sensor[xpos] = analogRead(A0);
  }
  // determine the peak to peak voltage
  maxvalue=sensor[1];
  minvalue=sensor[1];
  for (int xpos = 0; xpos <160; xpos++){
```

```
    if (sensor[xpos] > maxvalue) maxvalue=sensor[xpos];
    if (sensor[xpos] < minvalue) minvalue=sensor[xpos];
  }
  value=maxvalue-minvalue;
  // erase the screen to start again
  TFTscreen.background(0, 0, 0);
  // display the collected data
  for (int xpos = 0; xpos <159; xpos++){
  // select the color = B,G,R
  TFTscreen.stroke(50, 50, 255);
  // draw the line (xPos1, yPos1, xPos2, yPos2);
  TFTscreen.line(xpos, sensor[xpos]/8, xpos+1, sensor[xpos+1]/8);
  // Set font color to green
  TFTscreen.stroke(0, 255, 0);
  // Write the text value of the sensor
  if (xpos==0) TFTscreen.text( itoa(value/2, buf, 10), 10, 110);
  if (xpos==0) TFTscreen.text( "mv", 50, 110);
  }
}
```

Now let's add an analog to digital converter on pins D0 to D7. In software the "PINC" parallel input command that was used in some other projects is now replaced with "PIND". Now we will switch to the TLC5510 analog to digital converter. The TLC5510 is an 8 bit fast analog to digital converter. It comes in a SOP package so you will have to use an adapter to plug it into a breadboard.

Note that the TLC5510 outputs now go to D0 to D7 of the Arduino. You will need to either disconnect D0 and D1 in order to update the software on the Arduino or use two 1K ohm resistors as shown in the TLC5510 schematic. These same D0 and D1 pins are always used for the Arduino to communicate over the USB connection. However this is the only fast way to get 8 bits in parallel into an Arduino Uno.

Now to put some of those free analog input pins to work. You can connect four momentary contact switches to A2, A3, A4 and A5 to select up, down, right and left. Up and down will select the item to change and right and left will change the values of the selected item. You could also put all four switches on one analog pin by having each one select a different voltage of a resistor divider.

Up next is the schematic of the switches.

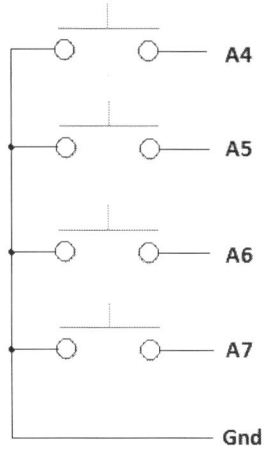

Last of all there are many software improvements. We can now switch between multiple selections like the delay between samples (as we did before), as well as the trigger level, and the trigger phase.

Another software improvement is that the trigger is now very "solid". It first checks for a positive going signal then for a negative going one. Otherwise it might just catch the positive level anywhere and proceed to collect samples. The trigger also has a timeout capability in case the trigger level never happens.

Here is a picture of the 1.8 inch SPI TFT LCD screen while running the PIND oscilloscope program.

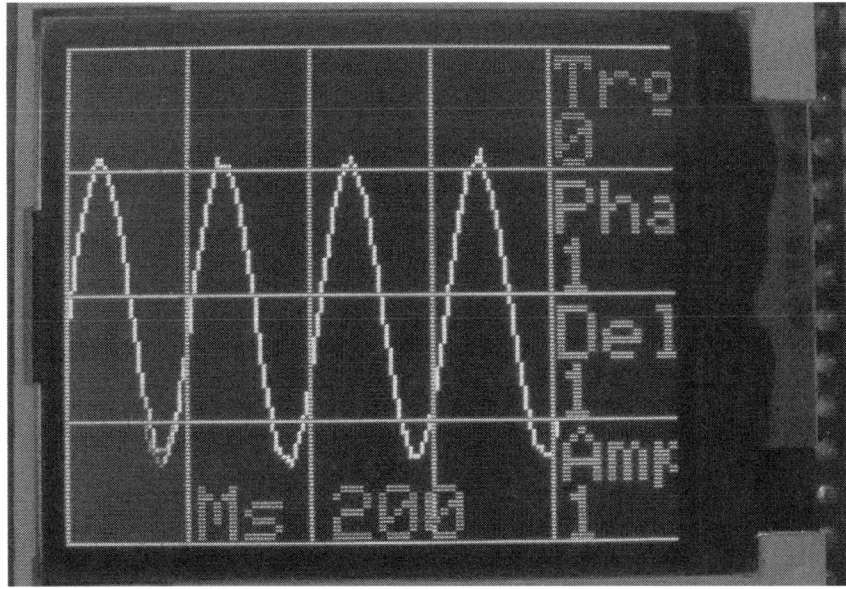

Here is the complete software listing.

```
/************************************
 1.8 SPI PIND TFT Oscilloscope
 Reads the D0-D7 pins using PIND,
 and shows the value on the screen.
 Created 7 July 2015 by Bob Davis
 *************************************/
#include <TFT.h>  // Arduino LCD library
#include <SPI.h>

// pin definition for the Uno LCD
#define rst  8
#define dc   9
#define cs   10
TFT TFTscreen = TFT(cs, dc, rst);

// set up variables
int xPos = 0;
char buf[12];
int Input=0;
byte Sample[160];
int StartSTime=0;
int EndSTime=0;
int STime=0;
int trigger=64;
int trigphase=1;
int tdelay=1;
int select=1;
int gain=1;

void setup(){
  // initialize the display
  TFTscreen.begin();
  // clear the screen
  TFTscreen.background(0, 0, 0);
  // Set the font size
  TFTscreen.setTextSize(2);
  // A to D input pins
  pinMode(0, INPUT);
  pinMode(1, INPUT);
  pinMode(2, INPUT);
```

```
  pinMode(3, INPUT);
  pinMode(4, INPUT);
  pinMode(5, INPUT);
  pinMode(6, INPUT);
  pinMode(7, INPUT);
  // Push button switches
  pinMode(16, INPUT_PULLUP);
  pinMode(17, INPUT_PULLUP);
  pinMode(18, INPUT_PULLUP);
  pinMode(19, INPUT_PULLUP);
}

void loop(){
  // wait for a positive going trigger
  if (trigphase==1){
    for (int timeout=0; timeout < 1000; timeout++){
      Input = PIND;
      if (Input < trigger) break;  }
    for (int timeout=0; timeout < 1000; timeout++){
      Input = PIND;
      if (Input > trigger) break;  }
    }
  // wait for a negative going trigger
  if (trigphase==0){
    for (int timeout=0; timeout < 1000; timeout++){
      Input = PIND;
      if (Input > trigger) break;  }
    for (int timeout=0; timeout < 1000; timeout++){
      Input = PIND;
      if (Input < trigger) break;  }
    }
  // quickly collect the data with no delay
  if (tdelay==-1){
    StartSTime = micros();
    Sample[0]=PIND;
    Sample[1]=PIND;     Sample[2]=PIND;     Sample[3]=PIND;
    Sample[4]=PIND;     Sample[5]=PIND;     Sample[6]=PIND;
    Sample[7]=PIND;     Sample[8]=PIND;     Sample[9]=PIND;
    Sample[10]=PIND;    Sample[11]=PIND;    Sample[12]=PIND;
    Sample[13]=PIND;    Sample[14]=PIND;    Sample[15]=PIND;
    Sample[16]=PIND;    Sample[17]=PIND;    Sample[18]=PIND;
    Sample[19]=PIND;    Sample[20]=PIND;    Sample[21]=PIND;
```

```
        Sample[22]=PIND;    Sample[23]=PIND;    Sample[24]=PIND;
        Sample[25]=PIND;    Sample[26]=PIND;    Sample[27]=PIND;
        Sample[28]=PIND;    Sample[29]=PIND;    Sample[30]=PIND;
        Sample[31]=PIND;    Sample[32]=PIND;    Sample[33]=PIND;
        Sample[34]=PIND;    Sample[35]=PIND;    Sample[36]=PIND;
        Sample[37]=PIND;    Sample[38]=PIND;    Sample[39]=PIND;
        Sample[40]=PIND;    Sample[41]=PIND;    Sample[42]=PIND;
        Sample[43]=PIND;    Sample[44]=PIND;    Sample[45]=PIND;
        Sample[46]=PIND;    Sample[47]=PIND;    Sample[48]=PIND;
        Sample[49]=PIND;    Sample[50]=PIND;    Sample[51]=PIND;
        Sample[52]=PIND;    Sample[53]=PIND;    Sample[54]=PIND;
        Sample[55]=PIND;    Sample[56]=PIND;    Sample[57]=PIND;
        Sample[58]=PIND;    Sample[59]=PIND;    Sample[60]=PIND;
        Sample[61]=PIND;    Sample[62]=PIND;    Sample[63]=PIND;
        Sample[64]=PIND;    Sample[65]=PIND;    Sample[66]=PIND;
        Sample[67]=PIND;    Sample[68]=PIND;    Sample[69]=PIND;
        Sample[70]=PIND;    Sample[71]=PIND;    Sample[72]=PIND;
        Sample[73]=PIND;    Sample[74]=PIND;    Sample[75]=PIND;
        Sample[76]=PIND;    Sample[77]=PIND;    Sample[78]=PIND;
        Sample[79]=PIND;    Sample[80]=PIND;    Sample[81]=PIND;
        Sample[82]=PIND;    Sample[83]=PIND;    Sample[84]=PIND;
        Sample[85]=PIND;    Sample[86]=PIND;    Sample[87]=PIND;
        Sample[88]=PIND;    Sample[89]=PIND;    Sample[90]=PIND;
        Sample[91]=PIND;    Sample[92]=PIND;    Sample[93]=PIND;
        Sample[94]=PIND;    Sample[95]=PIND;    Sample[96]=PIND;
        Sample[97]=PIND;    Sample[98]=PIND;    Sample[99]=PIND;
        Sample[100]=PIND;   Sample[101]=PIND;   Sample[102]=PIND;
        Sample[103]=PIND;   Sample[104]=PIND;   Sample[105]=PIND;
        Sample[106]=PIND;   Sample[107]=PIND;   Sample[108]=PIND;
        Sample[109]=PIND;   Sample[110]=PIND;   Sample[111]=PIND;
        Sample[112]=PIND;   Sample[113]=PIND;   Sample[114]=PIND;
        Sample[115]=PIND;   Sample[116]=PIND;   Sample[117]=PIND;
        Sample[118]=PIND;   Sample[119]=PIND;   Sample[120]=PIND;
        Sample[121]=PIND;   Sample[122]=PIND;   Sample[123]=PIND;
        Sample[124]=PIND;   Sample[125]=PIND;   Sample[126]=PIND;
        Sample[127]=PIND;   Sample[128]=PIND;   Sample[129]=PIND;
        Sample[130]=PIND;
        EndSTime = micros();
}
// Collect the data with a no delay
// It will not allow a delay of 0 so this is here to do that
if (tdelay ==0){
```

```
  StartSTime = micros();
  for (int xpos=0; xpos <130; xpos++){
    Sample[xpos]=PIND;
  }
  EndSTime = micros();
}
// Collect the data with a variable delay
if (tdelay > 0){
  StartSTime = micros();
  for (int xpos=0; xpos <130; xpos++){
    Sample[xpos]=PIND;
    delayMicroseconds(tdelay);
  }
  EndSTime = micros();
}
STime = EndSTime - StartSTime;
// display the collected data
for (int xpos = 0; xpos <160; xpos++){
  // select the color black = B,G,R
  TFTscreen.stroke(0, 0, 0);
  // erase the old line
  TFTscreen.line(xpos+1, 0, xpos+1, 160);
  // select the color white = B,G,R
  TFTscreen.stroke(255, 255, 255);
  // draw the trace line (xPos1, yPos1, xPos2, yPos2);0
  // if (xpos<128)TFTscreen.line(xpos, (Sample[xpos]/2), xpos+1, Sample[xpos+1]/2);
  if (xpos<128){
    if (gain==1){
      TFTscreen.line(xpos, (128-Sample[xpos]/2), xpos+1, 128-Sample[xpos+1]/2);
    }
    if (gain==2){
      TFTscreen.line(xpos, (192-Sample[xpos]*1), xpos+1, 192-Sample[xpos+1]*1);
    }
    if (gain==3){
      TFTscreen.line(xpos, (320-Sample[xpos]*2), xpos+1, 320-Sample[xpos+1]*2);
    }
    if (gain==4){
```

```
      TFTscreen.line(xpos, (448-Sample[xpos]*3), xpos+1, 448-Sample[xpos+1]*3);
    }
    if (gain==5){
      TFTscreen.line(xpos, (576-Sample[xpos]*4), xpos+1, 576-Sample[xpos+1]*4);
    }
  }

  // Set font color to green
  TFTscreen.stroke(0, 255, 0);
  // draw the green lines
  TFTscreen.line(xpos, 0, xpos+1, 0);
  TFTscreen.line(xpos, 32, xpos+1, 32);
  TFTscreen.line(xpos, 64, xpos+1, 64);
  TFTscreen.line(xpos, 96, xpos+1, 96);
  TFTscreen.line(xpos, 127, xpos+1, 127);
  // draw top to bottom green lines
  if (xpos==0) TFTscreen.line(xpos, 0, xpos,160);
  if (xpos==32) TFTscreen.line(xpos, 0, xpos,160);
  if (xpos==64) TFTscreen.line(xpos, 0, xpos,160);
  if (xpos==96) TFTscreen.line(xpos, 0, xpos,160);
  if (xpos==127) TFTscreen.line(xpos, 0, xpos,160);
}

// check the status of push button switches
// now circular so only two switches are required
if (digitalRead(17) == 0) select++;
if (digitalRead(16) == 0) select--;
if (select >4) select=1;
if (select <1) select=4;
// update the trigger level
if (select==1){
  if (digitalRead(18)==0) trigger++;
  if (digitalRead(19)==0) trigger--;
  if (trigger > 128) trigger=1;
  if (trigger < 1) trigger=128;
}
// change the trigger phase
if (select==2){
  if (digitalRead(18)==0) trigphase++;
  if (digitalRead(19)==0) trigphase--;
```

```
    if (trigphase < 0) trigphase=1;
    if (trigphase > 1) trigphase=0;
  }
  // Change the amount of delay
  if (select==3){
    // increase delay
    if (digitalRead(18)==0){
      if (tdelay==500) tdelay=-1;
      if (tdelay==200) tdelay=500;
      if (tdelay==100) tdelay=200;
      if (tdelay==50) tdelay=100;
      if (tdelay==20) tdelay=50;
      if (tdelay==10) tdelay=20;
      if (tdelay==5) tdelay=10;
      if (tdelay==2) tdelay=5;
      if (tdelay==1) tdelay=2;
      if (tdelay==0) tdelay=1;
      if (tdelay==-1) tdelay=0;
    }
    if (digitalRead(19)==0){
      // decrease delay
      if (tdelay==-1) tdelay=500;
      if (tdelay==0) tdelay=-1;
      if (tdelay==1) tdelay=0;
      if (tdelay==2) tdelay=1;
      if (tdelay==5) tdelay=2;
      if (tdelay==10) tdelay=5;
      if (tdelay==20) tdelay=10;
      if (tdelay==50) tdelay=20;
      if (tdelay==100) tdelay=50;
      if (tdelay==200) tdelay=100;
      if (tdelay==500) tdelay=200;
    }
  }
  // Change the amount of gain
  if (select==4){
    if (digitalRead(18)==0) gain++;
    if (digitalRead(19)==0) gain--;
    if (gain > 5) gain=1;
    if (gain < 1) gain=5;
  }
```

```
// Update the text on the right side
// Set font color to bright blue
TFTscreen.stroke(255, 100, 100);
// if selected set font color to red
if (select == 1)TFTscreen.stroke(0, 0, 255);
TFTscreen.text( "Trg", 129, 3);
TFTscreen.text( itoa(64-trigger, buf, 10), 129, 18);
// Set font color to bright blue
TFTscreen.stroke(255, 100, 100);
// if selected set font color to red
if (select == 2)TFTscreen.stroke(0, 0, 255);
TFTscreen.text( "Pha", 129, 35);
TFTscreen.text( itoa(trigphase, buf, 10), 129, 50);
// Set font color to bright blue
TFTscreen.stroke(255, 100, 100);
// if selected set color to red
if (select == 3)TFTscreen.stroke(0, 0, 255);
TFTscreen.text( "Del", 129, 67);
TFTscreen.text( itoa(tdelay, buf, 10), 129, 82);
// Set font color to bright blue
TFTscreen.stroke(255, 100, 100);
// if selected set font color to red
if (select == 4)TFTscreen.stroke(0, 0, 255);
TFTscreen.text( "Amp", 130, 98);
TFTscreen.text( itoa(gain, buf, 10), 130, 113);
// display the sample time
TFTscreen.stroke(0, 0, 255);
TFTscreen.text( "Ms", 35, 113);
TFTscreen.text( itoa(STime, buf, 10), 70, 113);
}
// End of program
```

Chapter 6

Parallel Color LCD TFT240_262K

When it comes to higher resolution LCD's there are many of them to choose from. They vary in size from 1.8 inches to 3.2 inches. Many even larger sizes are also available. How do you choose a LCD to work with? First, look at the number of pins on the LCD's connector. The larger ones have two rows of what looks like about 20 pins each because they are 16 bit devices. This will not plug into most breadboards, and there are too many pins for an Arduino UNO. However if you have an Arduino Mega or what is called a "Funduino", then you have enough pins to connect to the larger 16 bit LCD's.

There are also a number of LCD's available on eBay that only have a flexible flat ribbon cable going to them. If you do not have a jack that matches that flat ribbon cable, or an adapter circuit board, you cannot use this type of LCD either.

The one I picked is labeled "HY-TFT240_262K", it has a 18 pin connector for the LCD data. It also has connectors for the touch screen and for the optional on board memory module. We are just concerned with the 18 pin connector for now. Some of these LCD's have a jumper for 3.3 or five volt operation. Make sure that any jumpers are set properly before powering on the LCD.

For this project we will be using the "UTFT library". That stands for "Universal Thin Film Transistor" library. These LCD displays are referred to as "TFT" or "Thin Film Transistor" LCD displays. They feature three color operation as well as higher resolutions. They were designed for use in cell phones and MP4 players. The UTFT library is downloaded as "UTFT.rar". Because UTFT.rar is a "rar" file, and not a "zip" file, you might have to also download a "rar" compatible decompression program like "7-Zip" to decompress it. Once again, you place the decompressed files in your "Arduino\Libraries" directory. It should look like the picture below once it is properly installed.

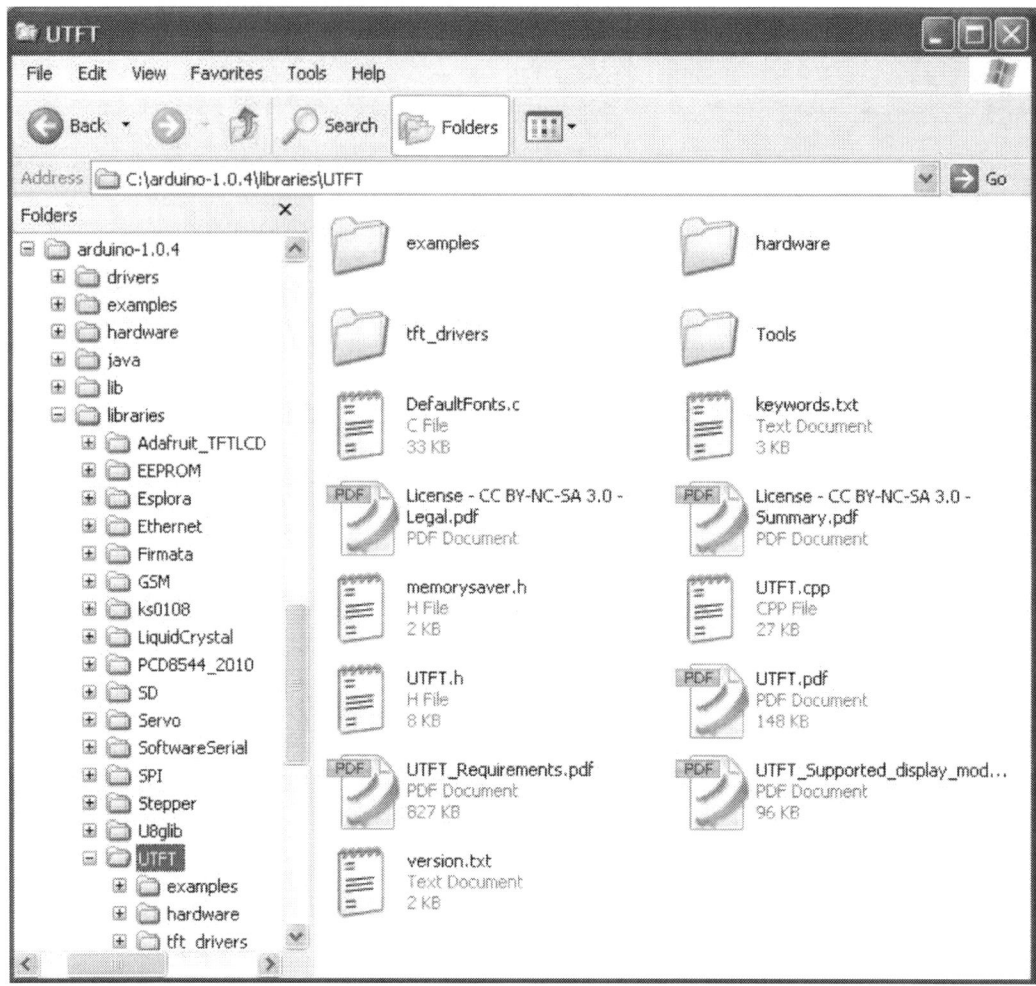

Here is a list of the UTFT Commands

UTFT (Model, RS, WR, CS, RST, ALE);
InitLCD(Orientation) ;   getDisplayXSize();
getDisplayYSize();   lcdOff();
lcdOn();   setContrast(c);
clrScr();   fillScr(r, g, b); fillScr(color);
setColor(r, g, b); setColor(color);
getColor();   setBackColor(r, g, b); setBackColor(color);
getBackColor();   drawPixel(x, y);
drawLine(x1, y1, x2, y2);   drawRect(x1, y1, x2, y2);
drawRoundRect(x1, y1, x2, y2);
fillRect(x1, y1, x2, y2);   fillRoundRect(x1, y1, x2, y2);
drawCircle(x, y, radius);   fillCircle(x, y, radius);

print(st, x, y, deg);   printNumI(num, x, y, length, filler);
printNumF(num, dec, x, y, divider, length, filler);
setFont(fontname);   getFont();
getFontXsize();   getFontYsize();

Here is a chart of some of the many chip sets that are supported by the UTFT library. Note that many of the supported chips are available in 8 bit, 16 bit and serial versions. The Arduino UNO can be used with 8 bit and serial versions.

Controller	Model for UTFT()	Supported mode		
		8bit	16bit	Serial
HX8340-B(N)	HX8340B_S			✓
HX8340-B(T)	HX8340B_8	✓		
HX8347-A	HX8347A		✓	
HX8352-A	HX8352A		✓	
ILI9320	ILI9320_8	✓		
	ILI9320_16		✓	
ILI9325C	ILI9325C	✓		
ILI9325D	ILI9325D_8	✓		
	ILI9325D_16		✓	
ILI9327	ILI9327		✓	
ILI9481	ILI9481		✓	
PCF8833	PCF8833			✓
S1D19122	S1D19122		✓	
S6D1121	S6D1121_8	✓		
	S6D1121_16		✓	
SSD1289	SSD1289		✓	
	SSD1289_8	✓		
	SSD1289LATCHED		LATCHED	
SSD1963	SSD1963_480		✓	
	SSD1963_800		✓	
	SSD1963_800ALT		✓	
ST7735	ST7735			✓

According to the UTFT library manual, this is how to connect the supported TFT LCD's to the Arduino processors. Note that DB0 to DB7 become DB8 to DB15 for 8 bit interfaces. This list also shows how to connect a 16 bit device to the Arduino UNO but that setup uses every last available pin. The TFT pin listing does not match our TFT display. The pinout listed is for 16 bit displays.

Signal	TFT pin	Arduino	
		2009/Uno/Leonardo	Mega/Due[2]
DB0[5]	21	D8	D37
DB1[5]	22	D9	D36
DB2[5]	23	D10	D35
DB3[5]	24	D11	D34
DB4[5]	25	D12	D33
DB5[5]	26	D13	D32
DB6[5]	27	A0 (D14)	D31
DB7[5]	28	A1 (D15)	D30
DB8	7	D0	D22
DB9	8	D1	D23
DB10	9	D2	D24
DB11	10	D3	D25
DB12	11	D4	D26
DB13	12	D5	D27
DB14	13	D6	D28
DB15	14	D7	D29
RS	4	Any free pin	
WR	5	Any free pin	
RD	6	Must be pulled high (3.3v)	
CS	15	Any free pin	
REST	17	Any free pin	

Here is another pinout chart for connecting the 18 pin LCD to the Arduino UNO.

```
LCD    Name      Arduino
------ -------   ---------
1      Gnd       Gnd
2      +5V       5V
3      NC
4      RS        A5
5      RW        A4
6      RD        3.3V
7      DB0       D0
8      DB1       D1
9      DB2       D2
10     DB3       D3
11     DB4       D4
12     DB5       D5
```

13	DB6	D6
14	DB7	D7
15	CS	A3
16	NC	
17	Rst	A2
18	NC	

Here is a schematic diagram of how to connect up the LCD screen.

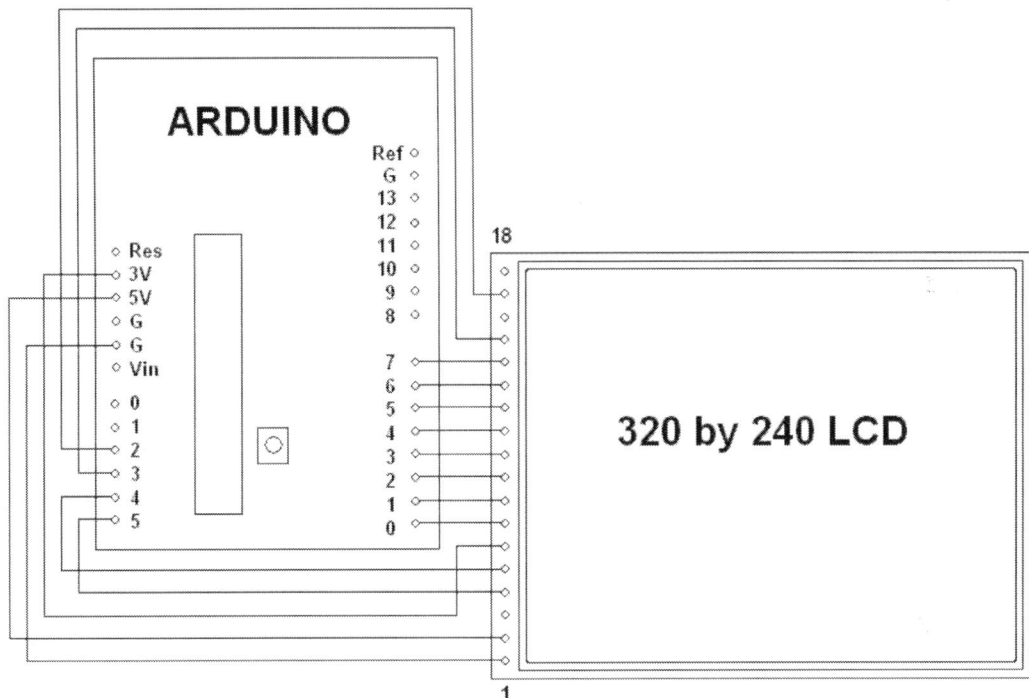

Up next is a picture of the oscilloscope program running. It had green division markers, blue text showing the millivolts peak to peak, and blue text showing the sample time. A switch connected to A1 can select fast sampling when ground, or slow sampling when it is at 5V. A software trigger loop makes the waveform fairly stable. For the analog input circuit schematic, see the oscilloscope input chapter earlier in this book.

Here is the sketch for the three color Arduino oscilloscope
//***
// Three color Fast analog Oscilloscope
// By Bob Davis
// UTFT_(C)2012 Henning Karlsen
// web: http://www.henningkarlsen.com/electronics

#include <UTFT.h>
#ifdef cbi
#define cbi(sfr, bit) (_SFR_BYTE(sfr) &= ~_BV(bit))
#endif
#ifdef sbi
#define sbi(sfr, bit) (_SFR_BYTE(sfr) |= _BV(bit))
#endif

// configure what fonts we will be using
extern uint8_t SmallFont[];
extern uint8_t BigFont[];
extern uint8_t SevenSegNumFont[];

UTFT myGLCD(ILI9325C,19,18,17,16);
int Input=0;
int OldInput=0;
int MaxSample=0;
int MinSample=0;
int Sample[320];
int OldSample[320];

```
int StartTime=0;
int EndTime=0;
int rate=1;

void DrawMarkers(){
  myGLCD.setColor(0, 200, 0);
  myGLCD.drawLine(0, 0, 0, 240);
  myGLCD.drawLine(54, 0, 54, 240);
  myGLCD.drawLine(107, 0, 107, 240);
  myGLCD.drawLine(160, 0, 160, 240);
  myGLCD.drawLine(213, 0, 213, 240);
  myGLCD.drawLine(266, 0, 266, 240);
  myGLCD.drawLine(319, 0, 319, 240);
  myGLCD.drawLine(0, 0, 319, 0);
  myGLCD.drawLine(0, 50, 319, 50);
  myGLCD.drawLine(0, 100, 319, 100);
  myGLCD.drawLine(0, 150, 319, 150);
  myGLCD.drawLine(0, 200, 319, 200);
  myGLCD.drawLine(0, 239, 319, 239);
}

void setup() {
  myGLCD.InitLCD();
  myGLCD.clrScr();
  myGLCD.setBackColor(0, 0, 0);
  myGLCD.setFont(BigFont);
}

void loop() {
  // set color(Red, Green, Blue) range = 0 to 255
  char buf[12];
  while(1) {
  // Set sample speed according to switch on A1
  if (analogRead(A1) < 500){
    cbi(ADCSRA, ADPS2);
    rate=0;
  }
  else {
    sbi(ADCSRA, ADPS2);
    rate=1;
  }
  DrawMarkers();
```

```
// wait for a trigger of a positive going input
OldInput=analogRead(A0);
Input=analogRead(A0);
while (Input < OldInput){
  Input=analogRead(A0);
}
// collect the analog data into an array
// do not do division here, it makes it slower!
StartTime = micros();
for(int xpos=0; xpos<319; xpos++) {
  Sample[xpos]=analogRead(A0);
}
EndTime = micros();
// display the collected analog data from array
// Sample/4.1 because 1024 becomes 250 = 5 volts
for(int xpos=0; xpos<319; xpos++) {
  myGLCD.setColor(0, 0, 0);
  myGLCD.drawLine (xpos, OldSample[xpos]/4.1, xpos+1, OldSample[xpos+1]/4.1);
  myGLCD.setColor(255, 255, 255);
  myGLCD.drawLine (xpos, Sample[xpos]/4.1, xpos+1, Sample[xpos+1]/4.1);
}
// Determine sample voltage peak to peak
MaxSample = Sample[100];
MinSample = Sample[100];
for(int xpos=0; xpos<319; xpos++) {
  OldSample[xpos] = Sample[xpos];
  if (Sample[xpos] > MaxSample) MaxSample=Sample[xpos];
  if (Sample[xpos] < MinSample) MinSample=Sample[xpos];
}
//  Display sample voltage and time
myGLCD.setColor(0, 0, 255);
int SampleSize=MaxSample-MinSample;
myGLCD.print("MV=", 1, 220);
myGLCD.print(itoa(SampleSize, buf, 10), 44, 220);
int SampleTime=EndTime-StartTime;
myGLCD.print("uS=      ", 161, 220);
// Adjust time according to sample speed
if (rate==1) {
  SampleTime=(EndTime/1000-StartTime/1000);
  myGLCD.print("mS=      ", 161, 220);
}
```

myGLCD.print(itoa(SampleTime, buf, 10), 210, 220);
  }
}

For the next project I will also show you how to make a six channel logic analyzer using this LCD display. You can use a loop to gather the data via "PINC" at about 1.33 million samples per second. You could also say "Sample[1]=PINC, Sample[2]=PINC, Sample[3]=PINC" etc, and then you can reach a speed of about 5 million samples per second.

You can also remove the Boolean statements under "// draw new data" and instead say "myGLCD.drawLine (xpos, Sample[xpos], xpos+1, Sample[xpos+1])" and you will then have an oscilloscope when a fast "flash" type of external analog to digital converter is connected.

Up next is the modified schematic with the changes needed to free up all of the analog input pins. Basically the wires going to A2-A5 are moved over to D8-D11.

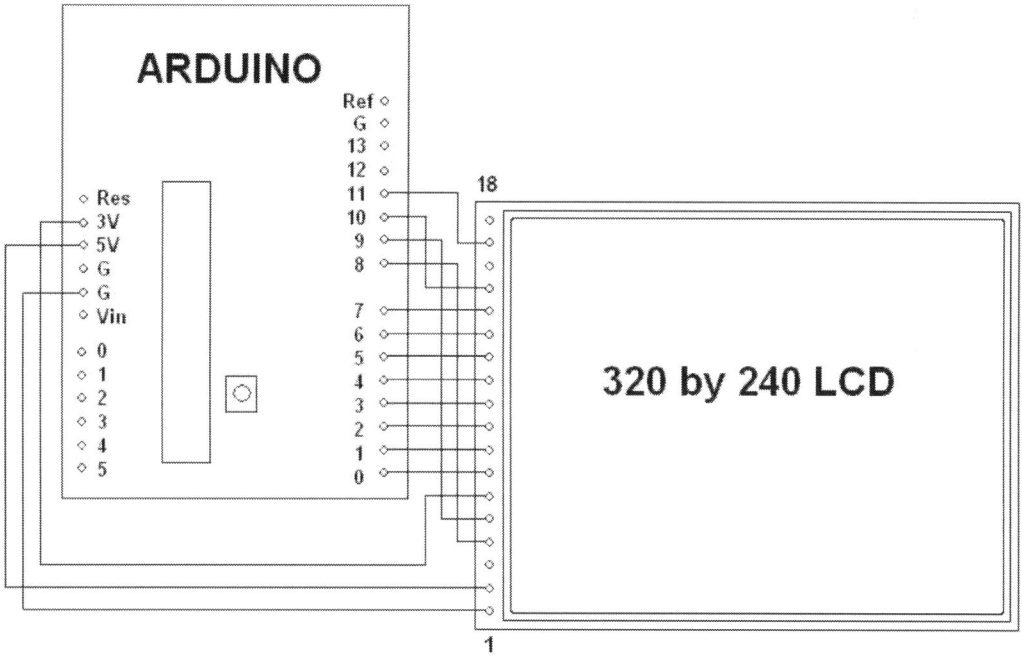

Up next is a picture of the six channel logic analyzer. It is showing six of the outputs of an AD775 fast analog to digital converter with a 20 KC sine wave on its input.

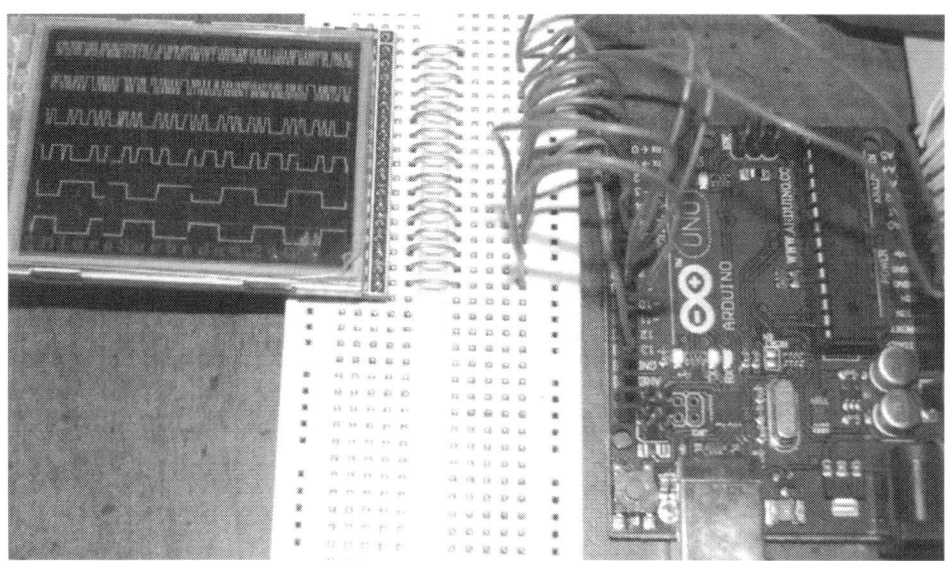

Here is the code for a sketch to make the logic analyzer work.

```
//*****************************************
// Three color 6 channel Logic Analyzer
// By Bob Davis
// UTFT_(C)2012 Henning Karlsen
// web: http://www.henningkarlsen.com/electronics

#include <UTFT.h>
// Declare which fonts we will be using
extern uint8_t SmallFont[];
extern uint8_t BigFont[];
extern uint8_t SevenSegNumFont[];

// Note that control pins are now assigned to pins D8-D11
UTFT myGLCD(ILI9325C,8,9,10,11);
int Input=0;
int Sample[320];
int StartSample=0;
int EndSample=0;

void DrawMarkers(){
  myGLCD.setColor(0, 200, 0);
  myGLCD.drawLine(0, 0, 0, 240);
  myGLCD.drawLine(54, 0, 54, 240);
  myGLCD.drawLine(107, 0, 107, 240);
```

```
  myGLCD.drawLine(160, 0, 160, 240);
  myGLCD.drawLine(213, 0, 213, 240);
  myGLCD.drawLine(266, 0, 266, 240);
  myGLCD.drawLine(319, 0, 319, 240);
  myGLCD.drawLine(0, 0, 319, 0);
  myGLCD.drawLine(0, 50, 319, 50);
  myGLCD.drawLine(0, 100, 319, 100);
  myGLCD.drawLine(0, 150, 319, 150);
  myGLCD.drawLine(0, 200, 319, 200);
  myGLCD.drawLine(0, 239, 319, 239);
}
void setup() {
  myGLCD.InitLCD();
  myGLCD.clrScr();
  pinMode(14, INPUT);
  pinMode(15, INPUT);
  pinMode(16, INPUT);
  pinMode(17, INPUT);
  pinMode(18, INPUT);
  pinMode(19, INPUT);
}
void loop() {
  // set color(Red, Green, Blue) range = 0 to 255
  myGLCD.setBackColor(0, 0, 0);
  myGLCD.setFont(BigFont);
  char buf[12];
  while(1) {
  DrawMarkers();
  // wait for trigger of a positive input
  while (Input == 0){
    Input=digitalRead(A0);
    }
// collect the analog data into an array
// Read analog port as a parallel port PINC
  StartSample = micros();
  for(int xpos=0; xpos<319; xpos++) {
    Sample[xpos]=PINC;
  }
  EndSample = micros();
// display the collected analog data from array
  for(int xpos=0; xpos<319; xpos++) {
    // Erase old stuff
```

```
  myGLCD.setColor(0, 0, 0);
  myGLCD.drawLine (xpos+1, 1, xpos+1, 220);
  // Draw new data
  myGLCD.setColor(255, 255, 255);
  myGLCD.drawLine (xpos, ((Sample[xpos]&B00000001)*16)+2, xpos+1,
((Sample[xpos+1]&B00000001)*16)+2);
  myGLCD.drawLine (xpos, ((Sample[xpos]&B00000010)*8)+42, xpos+1,
((Sample[xpos+1]&B00000010)*8)+42);
  myGLCD.drawLine (xpos, ((Sample[xpos]&B00000100)*4)+82, xpos+1,
((Sample[xpos+1]&B00000100)*4)+82);
  myGLCD.drawLine (xpos, ((Sample[xpos]&B00001000)*2)+122, xpos+1,
((Sample[xpos+1]&B00001000)*2)+122);
  myGLCD.drawLine (xpos, ((Sample[xpos]&B00010000)/1)+162, xpos+1,
((Sample[xpos+1]&B00010000)/1)+162);
  myGLCD.drawLine (xpos, ((Sample[xpos]&B00100000)/2)+202, xpos+1,
((Sample[xpos+1]&B00100000)/2)+202);
  }
  // display the sample time
  myGLCD.setColor(0, 0, 255);
  int SampleTime=EndSample-StartSample;
  myGLCD.print("MicroSeconds=", 10, 220);
  myGLCD.print(itoa(SampleTime, buf, 10), 224, 220);
    }
}
```

We can even make a faster oscilloscope if we add an external analog to digital converter such as the CA3306. It is a "flash" converter meaning that it has 64 comparators that instantly converts the analog input to a digital output. Because of that, it can do over 15 million conversions per second. A schematic of the CA3306 can be found in an earlier chapter in this book.

Here is the sketch code to make the fast oscilloscope work. Note that two momentary contact push buttons are added. One goes to D12 and one to D13, their other end connect to the ground. The switches select the "scan rate" and the "trigger level".

```
//***************************************
// Three color 5msps external AtoD Scope
// By Bob Davis
// UTFT_(C)2012 Henning Karlsen
// web: http://www.henningkarlsen.com/electronics
// Switches on D12 & D13 determine sweep speed and trigger level
```

//***

```cpp
#include <UTFT.h>
// Declare which fonts we will be using
extern uint8_t SmallFont[];
extern uint8_t BigFont[];
extern uint8_t SevenSegNumFont[];

// Note that the control pins are now assigned to 8-11
UTFT myGLCD(ILI9325C,8,9,10,11);
int Input=0;
byte Sample[320];
byte OldSample[320];
int StartSample=0;
int EndSample=0;
int MaxSample=0;
int MinSample=0;
int mode=0;
int dTime=1;
int Trigger=10;
int SampleSize=0;
int SampleTime=0;

void DrawMarkers(){
  myGLCD.setColor(0, 220, 0);
  myGLCD.drawLine(0, 0, 0, 240);
  myGLCD.drawLine(60, 0, 60, 240);
  myGLCD.drawLine(120, 0, 120, 240);
  myGLCD.drawLine(180, 0, 180, 240);
  myGLCD.drawLine(239, 0, 239, 240);
  myGLCD.drawLine(319, 0, 319, 240);
  myGLCD.drawLine(0, 0, 319, 0);
  myGLCD.drawLine(0, 60, 319, 60);
  myGLCD.drawLine(0, 120, 319, 120);
  myGLCD.drawLine(0, 180, 319, 180);
  myGLCD.drawLine(0, 239, 319, 239);
}

void setup() {
  myGLCD.InitLCD();
  myGLCD.clrScr();
  pinMode(12, INPUT);
```

```
  digitalWrite(12, HIGH);
  pinMode(13, INPUT);
  digitalWrite(13, HIGH);
  pinMode(14, INPUT);
  pinMode(15, INPUT);
  pinMode(16, INPUT);
  pinMode(17, INPUT);
  pinMode(18, INPUT);
  pinMode(19, INPUT);
}

void loop() {
// Set the background color(Red, Green, Blue)
  myGLCD.setBackColor(0, 0, 0);
  myGLCD.setFont(BigFont);
  char buf[12];
  while(1) {
    DrawMarkers();
    if (digitalRead(13) == 0) mode++;
    if (mode > 10) mode=0;
// Select delay times for scan modes
    if (mode == 0) dTime=0;
    if (mode == 1) dTime=0;
    if (mode == 2) dTime=1;
    if (mode == 3) dTime=2;
    if (mode == 4) dTime=5;
    if (mode == 5) dTime=10;
    if (mode == 6) dTime=20;
    if (mode == 7) dTime=50;
    if (mode == 8) dTime=100;
    if (mode == 9) dTime=200;
    if (mode == 10) dTime=500;
// Select trigger level
    if (digitalRead(12) == 0) Trigger=Trigger+10;
    if (Trigger > 50) Trigger=0;
// Wait for input to be greater than trigger
    while (Input < Trigger){
      Input = PINC;
    }

// Collect the analog data into an array
    if (mode == 0) {
```

```
// Read analog port as a parallel port no loop
  StartSample = micros();
  Sample[0]=PINC;
  Sample[1]=PINC;     Sample[2]=PINC;     Sample[3]=PINC;
  Sample[4]=PINC;     Sample[5]=PINC;     Sample[6]=PINC;
  Sample[7]=PINC;     Sample[8]=PINC;     Sample[9]=PINC;
  Sample[10]=PINC;    Sample[11]=PINC;    Sample[12]=PINC;
  Sample[13]=PINC;    Sample[14]=PINC;    Sample[15]=PINC;
  Sample[16]=PINC;    Sample[17]=PINC;    Sample[18]=PINC;
  Sample[19]=PINC;    Sample[20]=PINC;    Sample[21]=PINC;
  Sample[22]=PINC;    Sample[23]=PINC;    Sample[24]=PINC;
  Sample[25]=PINC;    Sample[26]=PINC;    Sample[27]=PINC;
  Sample[28]=PINC;    Sample[29]=PINC;    Sample[30]=PINC;
  Sample[31]=PINC;    Sample[32]=PINC;    Sample[33]=PINC;
  Sample[34]=PINC;    Sample[35]=PINC;    Sample[36]=PINC;
  Sample[37]=PINC;    Sample[38]=PINC;    Sample[39]=PINC;
  Sample[40]=PINC;    Sample[41]=PINC;    Sample[42]=PINC;
  Sample[43]=PINC;    Sample[44]=PINC;    Sample[45]=PINC;
  Sample[46]=PINC;    Sample[47]=PINC;    Sample[48]=PINC;
  Sample[49]=PINC;    Sample[50]=PINC;    Sample[51]=PINC;
  Sample[52]=PINC;    Sample[53]=PINC;    Sample[54]=PINC;
  Sample[55]=PINC;    Sample[56]=PINC;    Sample[57]=PINC;
  Sample[58]=PINC;    Sample[59]=PINC;    Sample[60]=PINC;
  Sample[61]=PINC;    Sample[62]=PINC;    Sample[63]=PINC;
  Sample[64]=PINC;    Sample[65]=PINC;    Sample[66]=PINC;
  Sample[67]=PINC;    Sample[68]=PINC;    Sample[69]=PINC;
  Sample[70]=PINC;    Sample[71]=PINC;    Sample[72]=PINC;
  Sample[73]=PINC;    Sample[74]=PINC;    Sample[75]=PINC;
  Sample[76]=PINC;    Sample[77]=PINC;    Sample[78]=PINC;
  Sample[79]=PINC;    Sample[80]=PINC;    Sample[81]=PINC;
  Sample[82]=PINC;    Sample[83]=PINC;    Sample[84]=PINC;
  Sample[85]=PINC;    Sample[86]=PINC;    Sample[87]=PINC;
  Sample[88]=PINC;    Sample[89]=PINC;    Sample[90]=PINC;
  Sample[91]=PINC;    Sample[92]=PINC;    Sample[93]=PINC;
  Sample[94]=PINC;    Sample[95]=PINC;    Sample[96]=PINC;
  Sample[97]=PINC;    Sample[98]=PINC;    Sample[99]=PINC;
  Sample[100]=PINC;   Sample[101]=PINC;   Sample[102]=PINC;
  Sample[103]=PINC;   Sample[104]=PINC;   Sample[105]=PINC;
  Sample[106]=PINC;   Sample[107]=PINC;   Sample[108]=PINC;
  Sample[109]=PINC;   Sample[110]=PINC;   Sample[111]=PINC;
  Sample[112]=PINC;   Sample[113]=PINC;   Sample[114]=PINC;
  Sample[115]=PINC;   Sample[116]=PINC;   Sample[117]=PINC;
```

```
Sample[118]=PINC;   Sample[119]=PINC;   Sample[120]=PINC;
Sample[121]=PINC;   Sample[122]=PINC;   Sample[123]=PINC;
Sample[124]=PINC;   Sample[125]=PINC;   Sample[126]=PINC;
Sample[127]=PINC;   Sample[128]=PINC;   Sample[129]=PINC;
Sample[130]=PINC;   Sample[131]=PINC;   Sample[132]=PINC;
Sample[133]=PINC;   Sample[134]=PINC;   Sample[135]=PINC;
Sample[136]=PINC;   Sample[137]=PINC;   Sample[138]=PINC;
Sample[139]=PINC;   Sample[140]=PINC;   Sample[141]=PINC;
Sample[142]=PINC;   Sample[143]=PINC;   Sample[144]=PINC;
Sample[145]=PINC;   Sample[146]=PINC;   Sample[147]=PINC;
Sample[148]=PINC;   Sample[149]=PINC;   Sample[150]=PINC;
Sample[151]=PINC;   Sample[152]=PINC;   Sample[153]=PINC;
Sample[154]=PINC;   Sample[155]=PINC;   Sample[156]=PINC;
Sample[157]=PINC;   Sample[158]=PINC;   Sample[159]=PINC;
Sample[160]=PINC;   Sample[161]=PINC;   Sample[162]=PINC;
Sample[163]=PINC;   Sample[164]=PINC;   Sample[165]=PINC;
Sample[166]=PINC;   Sample[167]=PINC;   Sample[168]=PINC;
Sample[169]=PINC;   Sample[170]=PINC;   Sample[171]=PINC;
Sample[172]=PINC;   Sample[173]=PINC;   Sample[174]=PINC;
Sample[175]=PINC;   Sample[176]=PINC;   Sample[177]=PINC;
Sample[178]=PINC;   Sample[179]=PINC;   Sample[180]=PINC;
Sample[181]=PINC;   Sample[182]=PINC;   Sample[183]=PINC;
Sample[184]=PINC;   Sample[185]=PINC;   Sample[186]=PINC;
Sample[187]=PINC;   Sample[188]=PINC;   Sample[189]=PINC;
Sample[190]=PINC;   Sample[191]=PINC;   Sample[192]=PINC;
Sample[193]=PINC;   Sample[194]=PINC;   Sample[195]=PINC;
Sample[196]=PINC;   Sample[197]=PINC;   Sample[198]=PINC;
Sample[199]=PINC;   Sample[200]=PINC;   Sample[201]=PINC;
Sample[202]=PINC;   Sample[203]=PINC;   Sample[204]=PINC;
Sample[205]=PINC;   Sample[206]=PINC;   Sample[207]=PINC;
Sample[208]=PINC;   Sample[209]=PINC;   Sample[210]=PINC;
Sample[211]=PINC;   Sample[212]=PINC;   Sample[213]=PINC;
Sample[214]=PINC;   Sample[215]=PINC;   Sample[216]=PINC;
Sample[217]=PINC;   Sample[218]=PINC;   Sample[219]=PINC;
Sample[220]=PINC;   Sample[221]=PINC;   Sample[222]=PINC;
Sample[223]=PINC;   Sample[224]=PINC;   Sample[225]=PINC;
Sample[226]=PINC;   Sample[227]=PINC;   Sample[228]=PINC;
Sample[229]=PINC;   Sample[230]=PINC;   Sample[231]=PINC;
Sample[232]=PINC;   Sample[233]=PINC;   Sample[234]=PINC;
Sample[235]=PINC;   Sample[236]=PINC;   Sample[237]=PINC;
Sample[238]=PINC;   Sample[239]=PINC;   Sample[240]=PINC;
EndSample = micros();
```

```
}
if (mode == 1) {
// Read analog port as a parallel port with no delay
  StartSample = micros();
  for(int xpos=0; xpos<240; xpos++) {
    Sample[xpos]=PINC;
  }
  EndSample = micros();
}
if (mode >= 2) {
// Read analog port as a parallel port variable delay
  StartSample = micros();
  for(int xpos=0; xpos<240; xpos++) {
    Sample[xpos]=PINC;
    delayMicroseconds(dTime);
  }
  EndSample = micros();
}

// Display the collected analog data from array
  for(int xpos=0; xpos<239; xpos++) {
// Erase the old stuff
    myGLCD.setColor(0, 0, 0);
    myGLCD.drawLine (xpos+1, 255-OldSample[xpos+1]*4, xpos+2, 255-OldSample[xpos+2]*4);
    if (xpos==0) myGLCD.drawLine (xpos+1, 1, xpos+1, 239);
// Draw the new data
    myGLCD.setColor(255, 255, 255);
    myGLCD.drawLine (xpos, 255-Sample[xpos]*4, xpos+1, 255-Sample[xpos+1]*4);
  }

// Determine sample voltage peak to peak
  MaxSample = Sample[100];
  MinSample = Sample[100];
  for(int xpos=0; xpos<240; xpos++) {
    OldSample[xpos] = Sample[xpos];
    if (Sample[xpos] > MaxSample) MaxSample=Sample[xpos];
    if (Sample[xpos] < MinSample) MinSample=Sample[xpos];
  }
// display the sample time, delay time and trigger level
  myGLCD.setColor(0, 0, 255);
```

```
    SampleTime=EndSample-StartSample;
    myGLCD.print("uSec.", 240, 10);
    myGLCD.print("    ", 240, 30);
    myGLCD.print(itoa(SampleTime, buf, 10), 240, 30);
    myGLCD.print("Delay", 240, 70);
    myGLCD.print("    ", 240, 90);
    myGLCD.print(itoa(dTime, buf, 10), 240, 90);
    myGLCD.print("Trig.", 240, 130);
    myGLCD.print(itoa(Trigger, buf, 10), 240, 150);
 // Range of 0 to 64 * 78 = 4992 mV
    SampleSize=(MaxSample-MinSample)*78;
    myGLCD.print("mVolt", 240, 190);
    myGLCD.print(itoa(SampleSize, buf, 10), 240, 210);
      }
  }
// end of program
```

Chapter 7

Serial Color LCD - 2.2 to 2.8 SPI

My next LCD for this book is a serial LCD based on the ILI9341 driver IC. These LCD screens are 240 by 320 pixels but still use a serial interface so they do not need to use a lot of the Arduino pins. Unfortunately these LCD's require either 3 volt or 2.5 volt logic. This logic level can easily be attained by using two 1K resistors to divide the 5 volt logic down to 2.5 volt logic. You can also use a 1K and a 2.2K resistor to get a 3 volt logic level. I tried using one 1K or 2.2K resistor in series but the LCD did not work that way.

The Adafruit version of this LCD screen has a 4050 IC on it that converts the logic level from 5 volts to 3 volts. There are also schematics available on the Internet that use a 4050 IC to interface the LCD.

When I first powered up the LCD screen the backlight LED was connected up to 3.3 volts. After a few minutes of operation the Arduino 3.3 volt regulator started shutting down. Instead it is best to connect the backlight LED to 5 volts via a 100 ohm resistor. In any case be sure to put a resistor in series with it, as there is no current limiting resistor built into the LCD.

Here are the Pin connections for this LCD.
1. VCC   Connect to 3.3 volts
2. Ground   Connect to ground
3. CS   Connect to D10 via resistor divider
4. Reset   Connect to 3.3 volts
5. D/C   Connect to D9 via resistor divider
6. MOSI/SDI   Connect to D11 via resistor divider
7. SCK   Connect to D13 via resistor divider
8. LED   Connect to 5V through 100 ohm resistor.
9. MISO   Not needed unless you have a touch screen.

Up next is the schematic diagram showing the 1k resistor dividers to provide 2.5 volt logic levels.

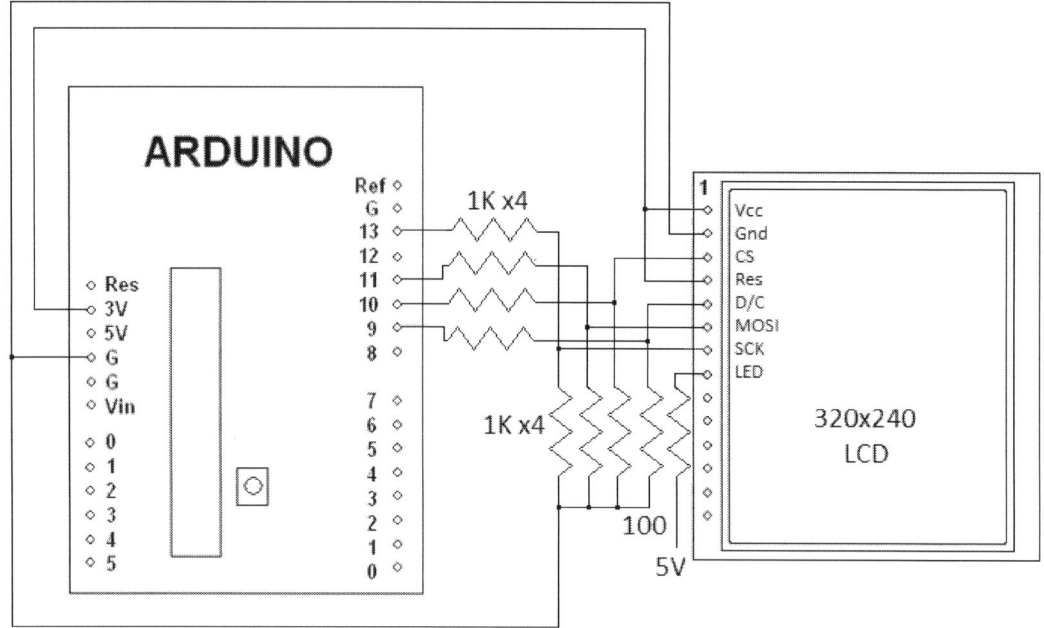

Adafruit has the best software drivers for use with this LCD. You will have to download, extract and then rename both the ILI9341 driver from Adafruit and the graphics (GFX) drivers from Adafruit.

Then even after doing that you may get an error with the "Robot Control" drivers whenever you try to upload your software to the Arduino. I know that does not make any sense, but if it does happen, then you can just rename the offending "Robot_Control" directory so that it will not be "valid". This will cause an error message when you start the Arduino interface, but you can just ignore that error message. You can see all of those changes to the Arduino libraries directory that are needed in the next picture.

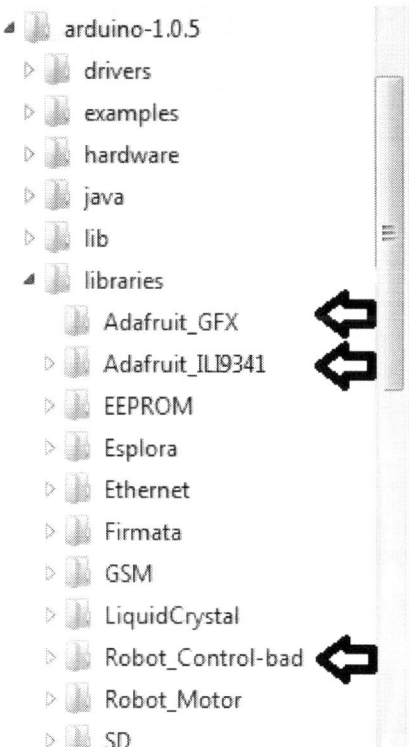

Also the Adafruit driver is a little bit on the slower side. Someone wrote a faster version but it goes under the exact same name so it is hard to tell them apart. The faster version is about twice as fast!

Here is a list of some of the commands this driver supports.
tft.drawLine (x1, y1, x2, y2, Color);
tft.drawPixel (x, y, Color);
tft.setCursor(x, y);;
tft.println("text");
tft.setTextColor(ILI9341_Color);
tft.setTextSize(2);
tft.begin();
tft.fillScreen(ILI9341_Color);
tft.setRotation(1);

Here is a list of the colors included with the driver.
RED	GREEN	BLUE	BLACK
YELLOW	WHITE	CYAN	BRIGHT_RED
GRAY1	GRAY2		

Here is a picture of a basic oscilloscope that is using the internal analog to digital converter running at 16 times its normal speed. The speedup routine causes slight inaccuracies that are visible at the top and bottom of the waveform. This is a 1000 Hz sine wave. Without the speedup routine a 100 Hz waveform will look almost identical, but it will be smoother.

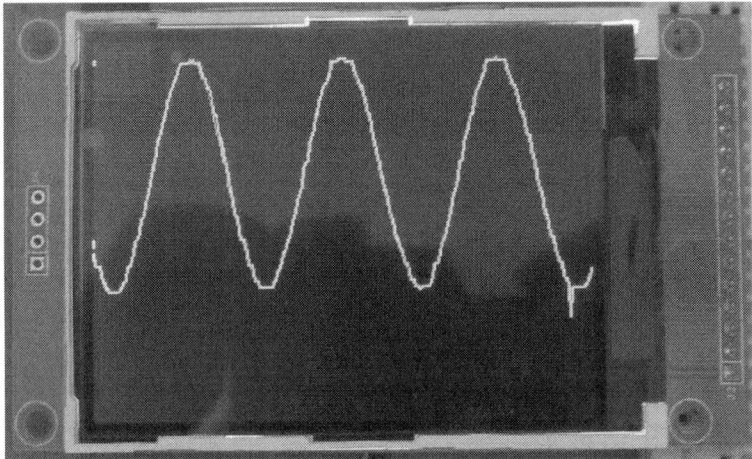

Here is a picture of a 100Hz sine wave with the converter running at its normal speed. Notice the smoothness of the sine wave.

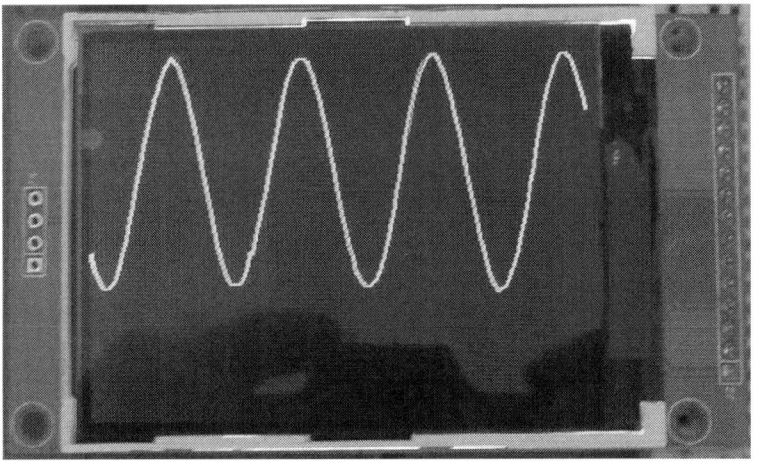

Here is the program listing for the simple oscilloscope using the internal analog to digital converter.

/************************************
2.8 SPI A0 TFT Oscope Simple
Reads the A0 analog input,

and shows the value on the screen.
Created 27 July 2015 by Bob Davis
*****************************/

```
#include <SPI.h>
#include "Adafruit_ILI9341.h"
#include "Adafruit_GFX.h"

// pin definition for the Uno LCD
#define TFT_DC 9
#define TFT_CS 10
// Use hardware SPI (on Uno, #13=clk, #11=mosi) and the above for CS/DC
//Adafruit_ILI9341 tft = Adafruit_ILI9341(TFT_CS, TFT_DC);
Adafruit_ILI9341 tft = Adafruit_ILI9341();

//// set up variables
int Input=0;
byte Sample[320];
int trigger=64;

void setup(){
  // initialize rotate and clear the display
  tft.begin();
  tft.fillScreen(ILI9341_BLACK);
  tft.setRotation(1);
  // Set the font size
  tft.setTextSize(2);
}

void loop(){
  // Speed up the converter.
  // Clear bit 2 of ADC pre-scalier from 125KHz to 2 MHz
  ADCSRA &= ~(1 << ADPS2);
  // wait for a positive going trigger
  for (int timeout=0; timeout < 1000; timeout++){
    Input = analogRead(A0);
    if (Input < trigger) break;  }
  for (int timeout=0; timeout < 1000; timeout++){
    Input = analogRead(A0);
    if (Input > trigger) break;  }
  // quickly collect the data with no delay
  for (int xpos=0; xpos <320; xpos++){
    Sample[xpos]=analogRead(A0);
```

```
}
// display the collected data
for (int xpos=0; xpos <319; xpos++){
  // erase the old and draw new line
  tft.drawLine(xpos+1, 0, xpos+1, 240, ILI9341_BLACK);
  tft.drawLine(xpos, (Sample[xpos]*2), xpos+1, Sample[xpos+1]*2, ILI9341_WHITE);
  }
}
// End of program
```

Here is a picture of the LCD working with an oscilloscope program that uses an external analog to digital converter.

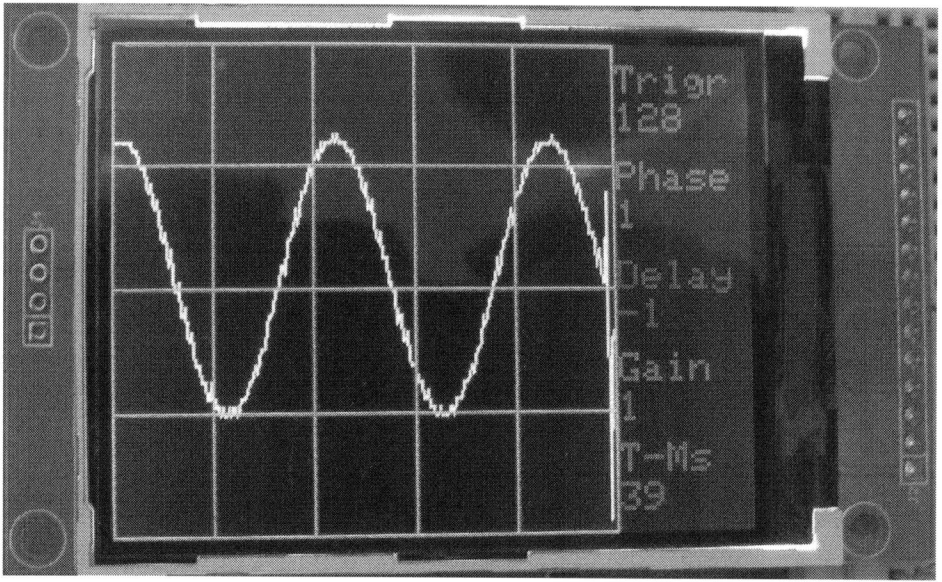

Here is the code for the oscilloscope with an external 8 bit analog to digital converter. This program also uses four momentary contact selection switches connected to A3-A5 with their common connection going to ground. The selected reading is highlighted in red, then the right and left buttons change the values.

```
/***********************************
2.8 SPI PIND TFT Oscilloscope
Reads the D0-D7 pins using PIND,
and shows the value on the screen.
Created 7 July 2015 by Bob Davis
************************************/
```

```
//#include <TFT.h>  // Arduino LCD library
#include <SPI.h>
#include "Adafruit_ILI9341.h"
#include "Adafruit_GFX.h"

// pin definitions for the ILI9341 2.8 LCD
#define TFT_DC 9
#define TFT_CS 10
// Use hardware SPI (on Uno, #13=clk, #12=nc, #11=mosi) and the above for CS/DC
// Adafruit_ILI9341 tft = Adafruit_ILI9341(TFT_CS, TFT_DC);
Adafruit_ILI9341 tft = Adafruit_ILI9341();

// set up the variables
int xPos=0;
char buf[12];
int Input=0;
byte Sample[250];
int StartSTime=0;
int EndSTime=0;
int stime=0;
int trigger=128;
int trigphase=1;
int tdelay=1;
int select=1;
int gain=1;

void setup(){
  // Initialize the display
  tft.begin();
  // Rotate and clear the screen
  tft.fillScreen(ILI9341_BLACK);
  tft.setRotation(1);
  // Set the font size
  tft.setTextSize(2);
  // A to D input pins
  pinMode(0, INPUT);
  pinMode(1, INPUT);
  pinMode(2, INPUT);
  pinMode(3, INPUT);
  pinMode(4, INPUT);
  pinMode(5, INPUT);
```

```
  pinMode(6, INPUT);
  pinMode(7, INPUT);
  // Push button switches
  pinMode(16, INPUT_PULLUP);
  pinMode(17, INPUT_PULLUP);
  pinMode(18, INPUT_PULLUP);
  pinMode(19, INPUT_PULLUP);
}

void loop(){
  // wait for a positive going trigger
  if (trigphase==1){
    for (int timeout=0; timeout < 1000; timeout++){
      Input = PIND;
      if (Input < trigger) break;  }
    for (int timeout=0; timeout < 1000; timeout++){
      Input = PIND;
      if (Input > trigger) break;  }
   }
  // wait for a negative going trigger
  if (trigphase==0){
    for (int timeout=0; timeout < 1000; timeout++){
      Input = PIND;
      if (Input > trigger) break;  }
    for (int timeout=0; timeout < 1000; timeout++){
      Input = PIND;
      if (Input < trigger) break;  }
   }
  // quickly collect the data with no delay
  if (tdelay==-1){
    StartSTime = micros();
    Sample[0]=PIND;
    Sample[1]=PIND;    Sample[2]=PIND;    Sample[3]=PIND;
    Sample[4]=PIND;    Sample[5]=PIND;    Sample[6]=PIND;
    Sample[7]=PIND;    Sample[8]=PIND;    Sample[9]=PIND;
    Sample[10]=PIND;   Sample[11]=PIND;   Sample[12]=PIND;
    Sample[13]=PIND;   Sample[14]=PIND;   Sample[15]=PIND;
    Sample[16]=PIND;   Sample[17]=PIND;   Sample[18]=PIND;
    Sample[19]=PIND;   Sample[20]=PIND;   Sample[21]=PIND;
    Sample[22]=PIND;   Sample[23]=PIND;   Sample[24]=PIND;
    Sample[25]=PIND;   Sample[26]=PIND;   Sample[27]=PIND;
    Sample[28]=PIND;   Sample[29]=PIND;   Sample[30]=PIND;
```

```
Sample[31]=PIND;   Sample[32]=PIND;   Sample[33]=PIND;
Sample[34]=PIND;   Sample[35]=PIND;   Sample[36]=PIND;
Sample[37]=PIND;   Sample[38]=PIND;   Sample[39]=PIND;
Sample[40]=PIND;   Sample[41]=PIND;   Sample[42]=PIND;
Sample[43]=PIND;   Sample[44]=PIND;   Sample[45]=PIND;
Sample[46]=PIND;   Sample[47]=PIND;   Sample[48]=PIND;
Sample[49]=PIND;   Sample[50]=PIND;   Sample[51]=PIND;
Sample[52]=PIND;   Sample[53]=PIND;   Sample[54]=PIND;
Sample[55]=PIND;   Sample[56]=PIND;   Sample[57]=PIND;
Sample[58]=PIND;   Sample[59]=PIND;   Sample[60]=PIND;
Sample[61]=PIND;   Sample[62]=PIND;   Sample[63]=PIND;
Sample[64]=PIND;   Sample[65]=PIND;   Sample[66]=PIND;
Sample[67]=PIND;   Sample[68]=PIND;   Sample[69]=PIND;
Sample[70]=PIND;   Sample[71]=PIND;   Sample[72]=PIND;
Sample[73]=PIND;   Sample[74]=PIND;   Sample[75]=PIND;
Sample[76]=PIND;   Sample[77]=PIND;   Sample[78]=PIND;
Sample[79]=PIND;   Sample[80]=PIND;   Sample[81]=PIND;
Sample[82]=PIND;   Sample[83]=PIND;   Sample[84]=PIND;
Sample[85]=PIND;   Sample[86]=PIND;   Sample[87]=PIND;
Sample[88]=PIND;   Sample[89]=PIND;   Sample[90]=PIND;
Sample[91]=PIND;   Sample[92]=PIND;   Sample[93]=PIND;
Sample[94]=PIND;   Sample[95]=PIND;   Sample[96]=PIND;
Sample[97]=PIND;   Sample[98]=PIND;   Sample[99]=PIND;
Sample[100]=PIND;  Sample[101]=PIND;  Sample[102]=PIND;
Sample[103]=PIND;  Sample[104]=PIND;  Sample[105]=PIND;
Sample[106]=PIND;  Sample[107]=PIND;  Sample[108]=PIND;
Sample[109]=PIND;  Sample[110]=PIND;  Sample[111]=PIND;
Sample[112]=PIND;  Sample[113]=PIND;  Sample[114]=PIND;
Sample[115]=PIND;  Sample[116]=PIND;  Sample[117]=PIND;
Sample[118]=PIND;  Sample[119]=PIND;  Sample[120]=PIND;
Sample[121]=PIND;  Sample[122]=PIND;  Sample[123]=PIND;
Sample[124]=PIND;  Sample[125]=PIND;  Sample[126]=PIND;
Sample[127]=PIND;  Sample[128]=PIND;  Sample[129]=PIND;
Sample[130]=PIND;  Sample[131]=PIND;  Sample[132]=PIND;
Sample[133]=PIND;  Sample[134]=PIND;  Sample[135]=PIND;
Sample[136]=PIND;  Sample[137]=PIND;  Sample[138]=PIND;
Sample[139]=PIND;  Sample[140]=PIND;  Sample[141]=PIND;
Sample[142]=PIND;  Sample[143]=PIND;  Sample[144]=PIND;
Sample[145]=PIND;  Sample[146]=PIND;  Sample[147]=PIND;
Sample[148]=PIND;  Sample[149]=PIND;  Sample[150]=PIND;
Sample[151]=PIND;  Sample[152]=PIND;  Sample[153]=PIND;
Sample[154]=PIND;  Sample[155]=PIND;  Sample[156]=PIND;
```

```
    Sample[157]=PIND;    Sample[158]=PIND;    Sample[159]=PIND;
    Sample[160]=PIND;    Sample[161]=PIND;    Sample[162]=PIND;
    Sample[163]=PIND;    Sample[164]=PIND;    Sample[165]=PIND;
    Sample[166]=PIND;    Sample[167]=PIND;    Sample[168]=PIND;
    Sample[169]=PIND;    Sample[170]=PIND;    Sample[171]=PIND;
    Sample[172]=PIND;    Sample[173]=PIND;    Sample[174]=PIND;
    Sample[175]=PIND;    Sample[176]=PIND;    Sample[177]=PIND;
    Sample[178]=PIND;    Sample[179]=PIND;    Sample[180]=PIND;
    Sample[181]=PIND;    Sample[182]=PIND;    Sample[183]=PIND;
    Sample[184]=PIND;    Sample[185]=PIND;    Sample[186]=PIND;
    Sample[187]=PIND;    Sample[188]=PIND;    Sample[189]=PIND;
    Sample[190]=PIND;    Sample[191]=PIND;    Sample[192]=PIND;
    Sample[193]=PIND;    Sample[194]=PIND;    Sample[195]=PIND;
    Sample[196]=PIND;    Sample[197]=PIND;    Sample[198]=PIND;
    Sample[199]=PIND;    Sample[200]=PIND;    Sample[201]=PIND;
    Sample[202]=PIND;    Sample[203]=PIND;    Sample[204]=PIND;
    Sample[205]=PIND;    Sample[206]=PIND;    Sample[207]=PIND;
    Sample[208]=PIND;    Sample[209]=PIND;    Sample[210]=PIND;
    Sample[211]=PIND;    Sample[212]=PIND;    Sample[213]=PIND;
    Sample[214]=PIND;    Sample[215]=PIND;    Sample[216]=PIND;
    Sample[217]=PIND;    Sample[218]=PIND;    Sample[219]=PIND;
    Sample[220]=PIND;    Sample[221]=PIND;    Sample[222]=PIND;
    Sample[223]=PIND;    Sample[224]=PIND;    Sample[225]=PIND;
    Sample[226]=PIND;    Sample[227]=PIND;    Sample[228]=PIND;
    Sample[229]=PIND;    Sample[230]=PIND;    Sample[231]=PIND;
    Sample[232]=PIND;    Sample[233]=PIND;    Sample[234]=PIND;
    Sample[235]=PIND;    Sample[236]=PIND;    Sample[237]=PIND;
    Sample[238]=PIND;    Sample[239]=PIND;    Sample[240]=PIND;
    Sample[241]=PIND;    Sample[242]=PIND;    Sample[243]=PIND;
    Sample[244]=PIND;    Sample[245]=PIND;    Sample[246]=PIND;
    Sample[247]=PIND;    Sample[248]=PIND;    Sample[249]=PIND;
    Sample[250]=PIND;
    EndSTime = micros();
}
// Collect the data with a no delay
// It will not allow a delay of 0 so this is here to do that
if (tdelay ==0){
  StartSTime = micros();
  for (int xpos=0; xpos <250; xpos++){
    Sample[xpos]=PIND;
  }
  EndSTime = micros();
```

```
}
// Collect the data with a variable delay
if (tdelay > 0){
  StartSTime = micros();
  for (int xpos=0; xpos <250; xpos++){
    Sample[xpos]=PIND;
    delayMicroseconds(tdelay);
  }
  EndSTime = micros();
}
stime = EndSTime - StartSTime;
// fix a bug in mode -1 that the displayed time is not correct
if (tdelay ==-1) stime = 49;
// display the collected data
for (int xpos = 0; xpos <320; xpos++){
  // erase the old line
  tft.drawLine(xpos+1, 0, xpos+1, 240, ILI9341_BLACK);
  // draw the trace line (xPos1, yPos1, xPos2, yPos2, color);
  // 256- inverts the data so it is right side up
  if (xpos<250){
    if (gain==1){
      tft.drawLine(xpos, (256-Sample[xpos]), xpos+1, 256-Sample[xpos+1], ILI9341_WHITE);
    }
    if (gain==2){
      tft.drawLine(xpos, (384-Sample[xpos]*2), xpos+1, 384-Sample[xpos+1]*2, ILI9341_WHITE);
    }
    if (gain==3){
      tft.drawLine(xpos, (512-Sample[xpos]*3), xpos+1, 512-Sample[xpos+1]*3, ILI9341_WHITE);
    }
    if (gain==4){
      tft.drawLine(xpos, (640-Sample[xpos]*4), xpos+1, 640-Sample[xpos+1]*4, ILI9341_WHITE);
    }
    if (gain==5){
      tft.drawLine(xpos, (768-Sample[xpos]*5), xpos+1, 768-Sample[xpos+1]*5, ILI9341_WHITE);
    }
  }
  // draw the horizontal green lines as dots
```

```
    if (xpos<250){
    tft.drawPixel(xpos, 0, ILI9341_GREEN);
    tft.drawPixel(xpos, 60, ILI9341_GREEN);
    tft.drawPixel(xpos, 120, ILI9341_GREEN);
    tft.drawPixel(xpos, 180, ILI9341_GREEN);
    tft.drawPixel(xpos, 239, ILI9341_GREEN);
    }
    // draw top to bottom green lines
    if (xpos==0) tft.drawLine(xpos, 0, xpos,240, ILI9341_GREEN);
    if (xpos==50) tft.drawLine(xpos, 0, xpos,240, ILI9341_GREEN);
    if (xpos==100) tft.drawLine(xpos, 0, xpos,240, ILI9341_GREEN);
    if (xpos==150) tft.drawLine(xpos, 0, xpos,240, ILI9341_GREEN);
    if (xpos==200) tft.drawLine(xpos, 0, xpos,240, ILI9341_GREEN);
    if (xpos==250) tft.drawLine(xpos, 0, xpos,240, ILI9341_GREEN);
}

//*** check the status of push button switches
// They are now circular, so only 2 switches are needed
if (digitalRead(17) == 0) select++;
if (digitalRead(16) == 0) select--;
if (select > 4) select=1;
if (select < 1) select=4;
// update the trigger level
if (select==1){
  if (digitalRead(18)==0) trigger++;
  if (digitalRead(19)==0) trigger--;
  if (trigger > 240) trigger=1;
  if (trigger < 1) trigger=240;
}
// change the trigger phase
if (select==2){
  if (digitalRead(18)==0) trigphase++;
  if (digitalRead(19)==0) trigphase--;
  if (trigphase < 1) trigphase=1;
  if (trigphase > 1) trigphase=0;
}
// Change the amount of delay
if (select==3){
  // increase delay
  if (digitalRead(18)==0){
    if (tdelay==500) tdelay=-1;
    if (tdelay==200) tdelay=500;
```

```
    if (tdelay==100) tdelay=200;
    if (tdelay==50) tdelay=100;
    if (tdelay==20) tdelay=50;
    if (tdelay==10) tdelay=20;
    if (tdelay==5) tdelay=10;
    if (tdelay==2) tdelay=5;
    if (tdelay==1) tdelay=2;
    if (tdelay==0) tdelay=1;
    if (tdelay==-1) tdelay=0;
  }
  if (digitalRead(19)==0){
    // decrease delay
    if (tdelay==-1) tdelay=500;
    if (tdelay==0) tdelay=-1;
    if (tdelay==1) tdelay=0;
    if (tdelay==2) tdelay=1;
    if (tdelay==5) tdelay=2;
    if (tdelay==10) tdelay=5;
    if (tdelay==20) tdelay=10;
    if (tdelay==50) tdelay=20;
    if (tdelay==100) tdelay=50;
    if (tdelay==200) tdelay=100;
    if (tdelay==500) tdelay=200;
  }
}
// Change the amount of gain
if (select==4){
  if (digitalRead(18)==0) gain++;
  if (digitalRead(19)==0) gain--;
  if (gain > 5) gain=1;
  if (gain < 1) gain=5;
}

//*** Update the text on the right side
tft.setTextColor(ILI9341_BLUE);
// if selected set font color to red
if (select == 1) tft.setTextColor(ILI9341_RED);
tft.setCursor(252, 10);
tft.println( "Trigr");
tft.setCursor(252, 29);
tft.println( itoa(128-trigger, buf, 10));
// if selected set font color to red
```

```
    tft.setTextColor(ILI9341_BLUE);
    // if selected set font color to red
    if (select == 2) tft.setTextColor(ILI9341_RED);
    tft.setCursor(252, 58);
    tft.println( "Phase");
    tft.setCursor(252, 77);
    tft.println( itoa(trigphase, buf, 10));
    // if selected set color to red
    tft.setTextColor(ILI9341_BLUE);
    // if selected set font color to red
    if (select == 3) tft.setTextColor(ILI9341_RED);
    tft.setCursor(252, 106);
    tft.println( "Delay");
    tft.setCursor(252, 125);
    tft.println( itoa(tdelay, buf, 10));
    tft.setTextColor(ILI9341_BLUE);
    // if selected set font color to red
    if (select == 4) tft.setTextColor(ILI9341_RED);
    tft.setCursor(252, 152);
    tft.println( "Gain");
    tft.setCursor(252, 171);
    tft.println( itoa(gain, buf, 10));
    tft.setTextColor(ILI9341_BLUE);
    // if selected set font color to red
    if (select == 5) tft.setTextColor(ILI9341_RED);
    tft.setCursor(252, 196);
    tft.println( "T-Ms");
    tft.setCursor(252, 215);
    tft.println( itoa(stime, buf, 10));
}
// End of program
```

Chapter 8

Putting the Oscilloscope into a Case

Once you come up with your favorite design it is time to put the Arduino powered oscilloscope into a case. I used a 6" by 4" by 2" plastic box from Radio Shack. I decided on the TLC5510 external analog to digital converter. Here is the layout of the plastic case. The LCD hole size shown is for a 2.8 inch LCD. You will need to adjust the hole size for your LCD screen size. The other holes are all 1/4 inches in diameter. Some switches might require slightly larger holes.

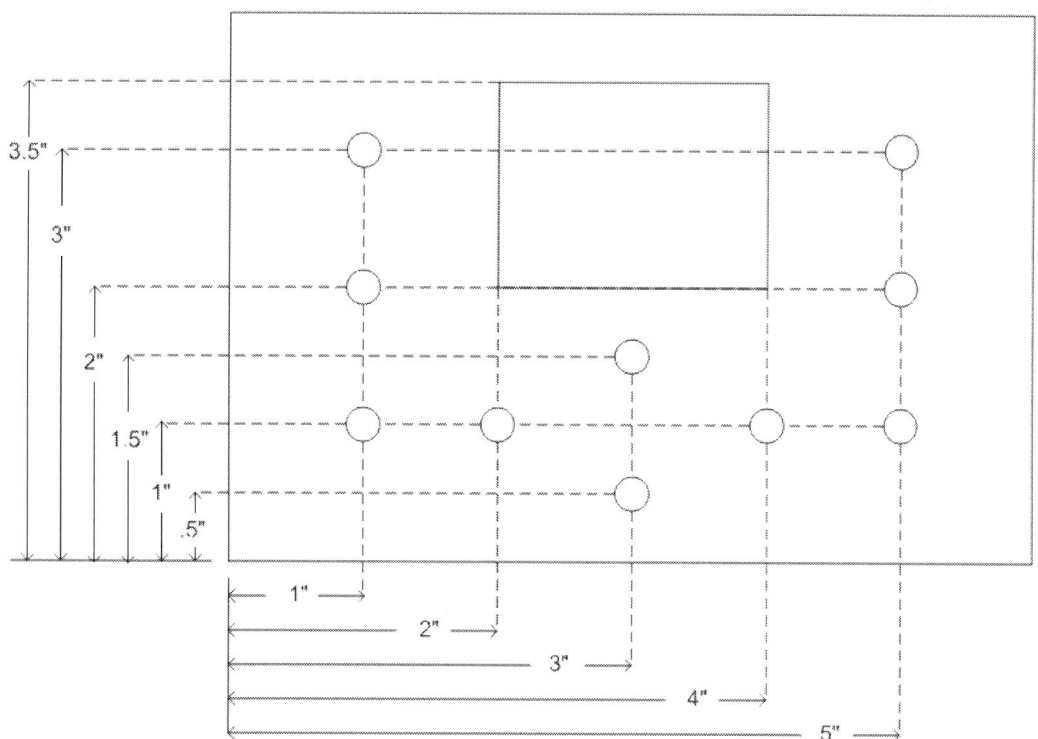

Here is a picture of the oscilloscope front panel controls with their labels.

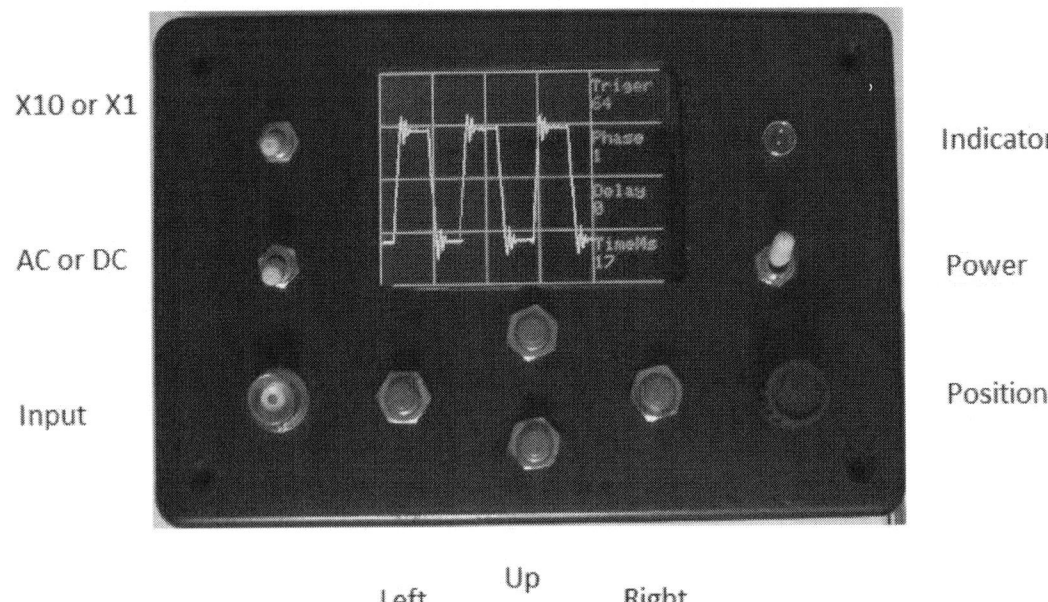

Try to layout the parts on the circuit board to make the wires as short and as simple as possible. Here is a picture of my parts layout. Notice the use of two 1/4 inch spacers to raise the LCD screen connector up a little higher.

The first thing you want to wire up will be all of the ground connections. They should be "redundant" as in they run in circles. The ground wires should be made with a larger wire like around 24-20 gauge wire.

Next wire up the power connections, they are +5 volts and +9 volts. They should also be a 20-24 gauge wire, but this time use insulated wire. Then wire the short direct connections that can be done with un-insulated wire like from the TLC5510 to the D0 to D7 pins. Also connect any pins that are to be to be shorted together. Then last of all go circuit by circuit and make sure that each one is completed.

Up next is a picture of the completed circuit board. It looks relatively simple. The front panel was not connected to the board when this picture was taken.

Here is a picture of the top side of the completed board. The wires on the left side go to the front panel of the oscilloscope.

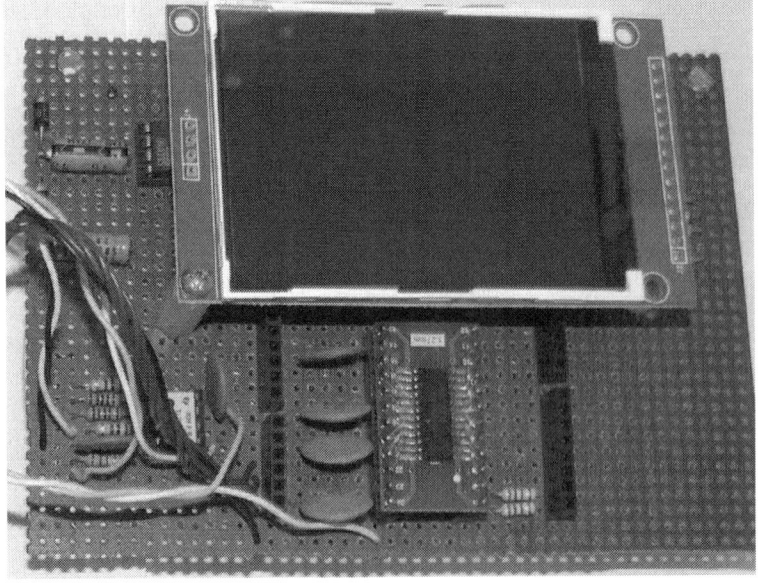

Bibliography

Programming Arduino
Getting Started With Sketches
By Simon Mark
Copyright 2012 by the McGraw-Hill Companies

This book gives a thorough explanation of the programming code for the Arduino. However the projects in the book are very basic. It does cover LCD's and Ethernet adapters.

Getting Started with Arduino
By Massimo Banzi
Copyright 2011 Massimo Banzi

This author is a co-founder of the Arduino.
This book has a quick reference to the programming code and some simple projects.

Arduino Cookbook
by Michael Margolis
Copyright © 2011 Michael Margolis and Nicholas Weldin. All rights reserved.
Printed in the United States of America.
Published by O'Reilly Media, Inc., 1005 Gravenstein Highway North, Sebastopol, CA.

This book has lots of great projects, with a very good explanation for every project.

Practical Arduino: Cool Projects for Open Source Hardware
Copyright © 2009 by Jonathan Oxer and Hugh Blemings
ISBN-13 (pbk): 978-1-4302-2477-8
ISBN-13 (electronic): 978-1-4302-2478-5
Printed and bound in the United States of America

Page 197 of this book has how to supercharge the Analog to Digital converter for faster sampling rates in oscilloscopes and logic analyzers.

Printed in Great Britain
by Amazon